European Conference on Process Safety and Big Data 2019

Frankfurt, Germany
1-3 October 2019

ISBN: 978-1-7138-2864-8

Printed from e-media with permission by:

Curran Associates, Inc.
57 Morehouse Lane
Red Hook, NY 12571

Some format issues inherent in the e-media version may also appear in this print version.

Printed with permission by Curran Associates, Inc. (2021)

For permission requests, please contact AIChE
at the address below.

AIChE
120 Wall Street, FL 23
New York, NY 10005-4020

Phone: (800) 242-4363
Fax: (203) 775-5177

www.aiche.org

Additional copies of this publication are available from:

Curran Associates, Inc.
57 Morehouse Lane
Red Hook, NY 12571 USA
Phone: 845-758-0400
Fax: 845-758-2633
Email: curran@proceedings.com
Web: www.proceedings.com

TABLE OF CONTENTS

INTRODUCTORY WELCOME ADDRESS

DAY 1 KEYNOTE SESSION

PANEL DISCUSSION: PROCESS INDUSTRY'S EVOLUTION TOWARDS BIG DATA FOR PROCESS SAFETY: CHALLENGES AND SUCCESSES

SESSION 1: TOOLS TO AVOID HUMAN ERRORS AND RAPID FIRE SESSION

SESSION 2: DATA LAKES / INTELLIGENT DASHBOARDS / KPIS

SESSION 3: DATA LAKES / INTELLIGENT DASHBOARDS / KPIS AND BIG DATA ANALYTICS

SESSION 4: BIG DATA ANALYTICS AND STANDARDS & METHODS

DAY 2 WELCOME ADDRESS

KEYNOTE 3

PANEL DISCUSSION: CURRENT TECHNOLOGY DEVELOPMENTS FOR EHS DIGITAL SOLUTIONS

SESSION 5: PRACTICAL EXAMPLES/CASE STUDIES AND PLANT MONITORING

SESSION 6: ROBOTICS IN SUPPORT OF SAFETY &PREDICTIVE MAINTENANCE

SESSION 7: STANDARD & METHODS AND AUGMENTED REALITY

2019 EUROPEAN CONFERENCE ON PROCESS SAFETY & BIG DATA

OCTOBER 1-2, 2019
DECHEMA HOUSE, FRANKFURT AM MAIN, GERMANY

CCPS EPSC

Welcome from CCPS & EPSC!

- Shakeel Kadri, Executive Director & CEO, CCPS
- Dr. Hans V. Schwarz, CEO, ProsafeX & EPSC Board Member

2019 EUROPEAN CONFERENCE ON PROCESS SAFETY & BIG DATA

OCTOBER 1-2, 2019
DECHEMA HOUSE, FRANKFURT AM MAIN, GERMANY

CCPS EPSC

2016
EUROPEAN CONFERENCE
ON PROCESS SAFETY AND
BIG DATA

2017
EUROPEAN CONFERENCE
ON PROCESS SAFETY AND
BIG DATA

2018 EUROPEAN CONFERENCE
ON PROCESS SAFETY & BIG DATA

CCPS EPSC

Our journey

2019 EUROPEAN CONFERENCE ON PROCESS SAFETY & BIG DATA

OCTOBER 1-2, 2019
DECHEMA HOUSE, FRANKFURT AM MAIN, GERMANY

CCPS EPSC

Thanks to Planning Committee Members:

1. Abdul Aldeeb - *Siemens*
2. Gus Carroll - *Centrica*
3. Ivan Castillo - *Dow Chemical*
4. Tony Downes - *Honeywell*
5. Arthur Groot - *DSM Resins & Functional Materials*
6. *Pol Hoorelbeke - TOTAL*
7. *Tijs Koerts - EPSC*
8. *Daniel Mottaghi – TOTAL*
9. Rendall Ricardo - Dow
10. Richard Pawulski - *Air Liquide*
11. Hans Schwarz – ProsafeX
12. Mike Snyder *DEKRA*

CCPS Staff:

Shakeel Kadri
CCPS Executive Director & CEO
Willi Meier
CCPS European Regional Manager
Jing Chen
CCPS Senior Engineering Specialist
Horst Massong
CCPS Europe Project Manager
Evan Flach
AIChE Associate Director, Sales

2019 EUROPEAN CONFERENCE ON PROCESS SAFETY & BIG DATA

OCTOBER 1-2, 2019
DECHEMA HOUSE, FRANKFURT AM MAIN, GERMANY

CCPS EPSC

THANK YOU TO OUR CONFERENCE SPONSORS

GOLD

AcuTech
PROCESS RISK MANAGEMENT

SILVER

EMERSON. evision SIEMENS Ingenuity for life

BRONZE

sphera

2019 EUROPEAN CONFERENCE ON PROCESS SAFETY & BIG DATA

OCTOBER 1-2, 2019
DECHEMA HOUSE, FRANKFURT AM MAIN, GERMANY

CCPS EPSC

THANK YOU TO OUR EXHIBITORS:

- AcuTech Consulting Group Inc
- CGE Risk Management Solutions
- Emerson
- eVision Industry Software
- Siemens Process & Safety Consulting
- Sphera
- Versatec Energy B.V.

2019 EUROPEAN CONFERENCE ON PROCESS SAFETY & BIG DATA

OCTOBER 1-2, 2019
DECHEMA HOUSE, FRANKFURT AM MAIN, GERMANY

CCPS EPSC

Thanks to All Presenters & Attendees!

- 106 Registrants as of September 30, 2019
- 55 Companies

2019 EUROPEAN CONFERENCE ON PROCESS SAFETY & BIG DATA

OCTOBER 1-2, 2019
DECHEMA HOUSE, FRANKFURT AM MAIN, GERMANY

CCPS EPSC

OCTOBER 01, 2019 – DAY 1	
8:30 AM - 8:50 AM	**WELCOME ADDRESS - Shakeel Kadri and Hans Schwarz**
8:50 AM - 10:00 AM Keynotes session chair: Shakeel Kadri	**KEYNOTE 1:** Integrating the Digital Field Worker **Torsten Schroeter, Benjamin Weiher,** and **Alex J. Wood,** Aveva
	KEYNOTE 2: Digital Tools to Help Operators and Technicians Avoid Incidents Caused by Human Error **Dror Barak,** GOARC
10:00 AM - 11:00 AM Panel session chair: Richard Pawulski	**PANEL DISCUSSION – Process Industry's Evolution Towards Big Data for Process Safety: Challenges and Successes** **Panelists:** • Dow: **Erik Jan van den Heuvel** • Dupont: **Nicolas Sands** • TOTAL: **Pol Hoorelbeke** • Shell: **Marijn Bezuijen**
11:00 AM - 11:25 AM	**Coffee Break**
11:25 AM - 12:55 PM	**SESSION 1: Tools to Avoid Human Errors**
Session chair: Willi Meier	Safety Through Digitization – Digital Field Solutions that Enable and Empower, **Sebastian Loeb,** Honeywell
	Decision Support Platforms and Industry 4.0, **Jim Petrusich,** Northwest Analytics
	Unprecedented Process Safety Improvements through Dynamic Barrier Risk Management, **Michel Tol,** eVision Industry Software
12:55 PM - 1:55 PM	**Lunch**

2019 EUROPEAN CONFERENCE ON PROCESS SAFETY & BIG DATA

OCTOBER 1-2, 2019
DECHEMA HOUSE, FRANKFURT AM MAIN, GERMANY

CCPS EPSC

1:55 PM - 3:25 PM	**SESSION 2: Data Lakes / Intelligent Dashboards / KPIs**
Session chair: Willi Meier	Overcoming of Accuracy and Computing Time of Online Model-Based Reactor Safety Concepts, **Johannes Biernath,** Center of Safety Excellence
	From Incidents to Intelligent KPIs: The Construction of Safety Indexes Based on Events, **Chabane Mazri, Ineris**
	Using an Artificial Intelligence Technique and Quantitative Data in Risk Assessment Matrix: Fuzzy Sets Approach **Pedro Prado,** Petronew Engenharia e Projetos and University of Campinas
3:25 PM - 3:50 PM	**Coffee Break**
3:50 PM - 5:20 PM	**SESSION 3: Data Lakes / Intelligent Dashboards / KPIs and Big Data Analytics**
Session Chair: Hans Schwarz	Integrating Operational and Enterprise Risk Management to Improve Asset Integrity and Reliability,**Sam Williams,** ERM and Igor Milic, MOL Group
	Big Data Analytics and Emergency Management, **Begum Dogan,** CASGEM
	Finding Health & Safety Buried Treasure with AI, **James Neophytou,** IBM
5:20 PM - 5:50 PM	**SESSION 4: Big Data Analytics and Standards**
Session chair: Pol Hoorelbeke	Applying Analytics to the Process Safety Life-Cycle: Reveal and Repair Process Safety Vulnerabilities, **Nicolas Sands,** DuPont and **Tony Downes,** Honeywell
5:50 PM	**RECEPTION & POSTER SESSION**

Note: Posters will be displayed during all breaks, Lunches and at Reception

2019 EUROPEAN CONFERENCE ON PROCESS SAFETY & BIG DATA

OCTOBER 1-2, 2019
DECHEMA HOUSE, FRANKFURT AM MAIN, GERMANY

CCPS EPSC

OCTOBER 02, 2019 – DAY 2	
8:30 AM - 8:50 AM	**WELCOME ADDRESS - Pol Hoorelbeke** and **Willi Meier**
8:50 AM - 9:25 AM Keynote session chair: Pol Hoorelbeke	**KEYNOTE 3:** Applying Safety Analytics for Improving Organizational Risk Capability to Reduce Incident Frequency and Severity: From Heavy Industry to the Modern Tech World, **Joe Stough**, WAITR, INC.
9:25 AM - 10:25 AM Panel session chair: Abdul Aldeeb & Tony Downes	**PANEL DISCUSSION - Current Technology Developments for EHS Digital Solutions** **Panelists:** • Sphera: **Uwe Schneider** • Honeywell: **Sebastian Loeb** • IBM: **Thomas Gervin**
10:25 AM - 10:50 AM	**Coffee Break**
10:50 AM - 12:20 PM	**SESSION 5: Practical Examples/Case Studies and Plant Monitoring**
Session chair: Daniel Mottaghi	Journey from a Disconnected to an Integrated Workflow in Process Plant Operation, **Sven Lohmann**, Emerson
	Cloud QRA: Developing Digital Twins for Offshore Installations, **Zoe Wattis**, DNV GL and MyQRA: Dynamic Quantitative Risk Assessment and Online Sensitivities, **Olivier Baldan**, DNV GL
	A Holistic Approach for Managing Past, Present and Future of Operational Process Safety, **Steve Elliott**, Schneider Electric
12:20 PM - 1:10 PM	**Lunch**

2019 EUROPEAN CONFERENCE ON PROCESS SAFETY & BIG DATA

OCTOBER 1-2, 2019
DECHEMA HOUSE, FRANKFURT AM MAIN, GERMANY

CCPS EPSC

1:10 PM - 2:40 PM	**SESSION 6: Robotics in Support of Safety & Predictive Maintenance**
Session chair: Mike Snyder	Insight in the Use of Drones at Total, **Julien da Vela**, TOTAL
	Equipment Predictive Analytics for Predictive Monitoring. **Monica Florentina Hildinger**, Siemens
	Applying Deep Learning and Sequence Mining in Alarm Management, **Janos Abonyi**, University of Pannonia
2:40 PM - 3:15 PM Keynote session chair: Pol Hoorelbeke	**KEYNOTE 4:** Artificial Intelligence for Process Safety: Past, Present, and Future, **Venkat Venkatasubramanian**, Columbia University (This presentation will be virtual)
3:15 PM - 3:40 PM	**Coffee Break**
3:40 PM - 4:40 PM	**SESSION 7: Standard & Methods and Augmented Reality**
Session chair: Hans Schwarz	Mission #2 to the Safety Data Universe , **Joep Coenen**, Versatec Energy and **Mark Van 't Zet**, Radial SG
	Is Big Data Analysis Really Worth for Instrumentation?, **Christian Demski**, Dow Chemical
4:40 PM	**Vote of Thanks & Plan for 2020**

Note: Posters will be displayed during all breaks, Lunches and at Reception

Conference Proceeding will be Available after the Conference:

www.aiche.org/ccpseuropeanconference

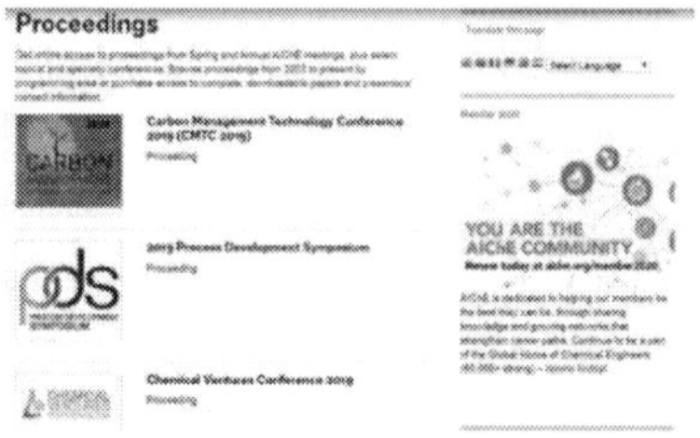

https://www.aiche.org/resources/proceedings

Thank You and Enjoy the Conference!

CCPS
Center for Chemical Process Safety
Safety 4.0™
Dror Barak CEO, GOARC - Safety 4.0
GOARC
Safety 4.0

Next Generation Safety
GOARC
AI -RISK PATTERN
RECOGNITION
SAFE CONNECTED
WORKER
Safety
4.0™
DYNAMIC™
ROUTINES
NEW
BEHAVIORAL
INSIGHTS

Chapter 1
Inspiration
GOARC

BBS VS HOP
Scott Geller & Todd Conklin
Safety 2017 - Plenary Town Hall(Tom Krause),

HOP
GOARC
"Our facilities are **NOT** inherently Safe. PEOPLE create safety in practice. "
TODD CONKLIN
Pre-Accident Investigations
BEHIND HUMAN ERROR

IMPLEMENTATION GAP
GOARC
INDUSTRY4.0 for PROCESS SAFETY HANDBOOK
JIM PETRUSICH
HANS VOLKMAR SCHWARZ
Safety Differently
Sidney Dekker
The Field Guide to Understanding 'Human Error'
Sidney Dekker
WORKPLACE FATALITIES
GUIDELINES FOR INVESTIGATING PROCESS SAFETY INCIDENTS
WILEY
Guidelines for Preventing Human Error in Process Safety
Pre-Accident Investigations
An Introduction to Organizational Safety
Todd Conklin
BEHIND HUMAN ERROR
SECOND EDITION
Process Safety Leadership from the Boardroom to the Frontline

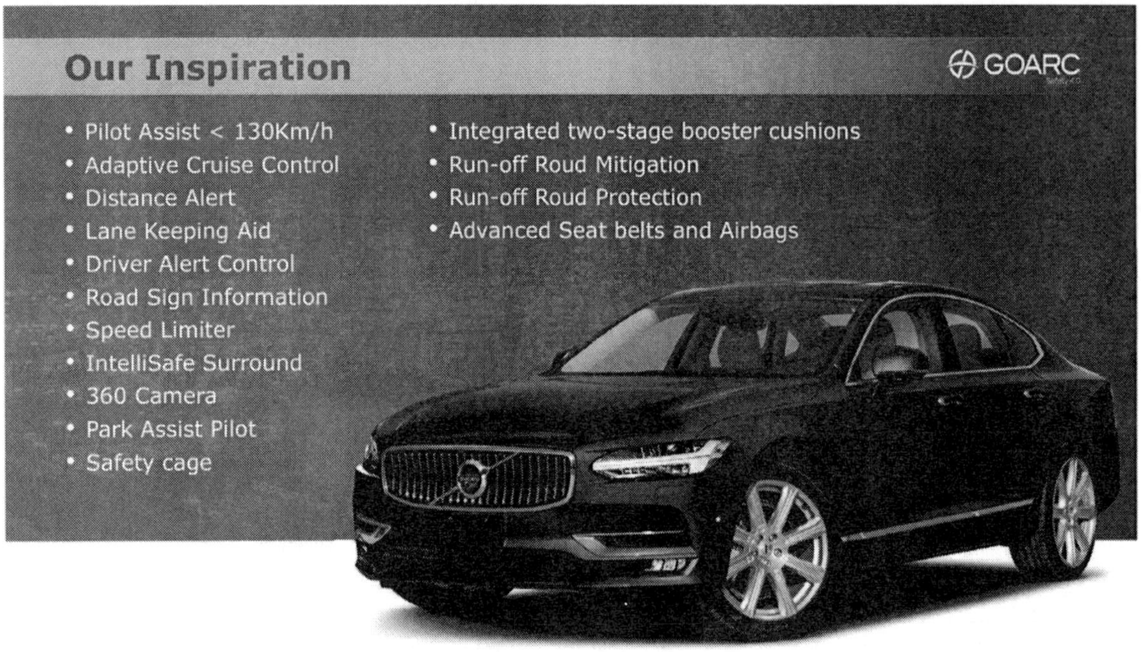
Our Inspiration
GOARC
• Pilot Assist < 130Km/h
• Adaptive Cruise Control
• Distance Alert
• Lane Keeping Aid
• Driver Alert Control
• Road Sign Information
• Speed Limiter
• IntelliSafe Surround
• 360 Camera
• Park Assist Pilot
• Safety cage
• Integrated two-stage booster cushions
• Run-off Roud Mitigation
• Run-off Roud Protection
• Advanced Seat belts and Airbags

Chapter 2
GOARC
The challenges
GOARC

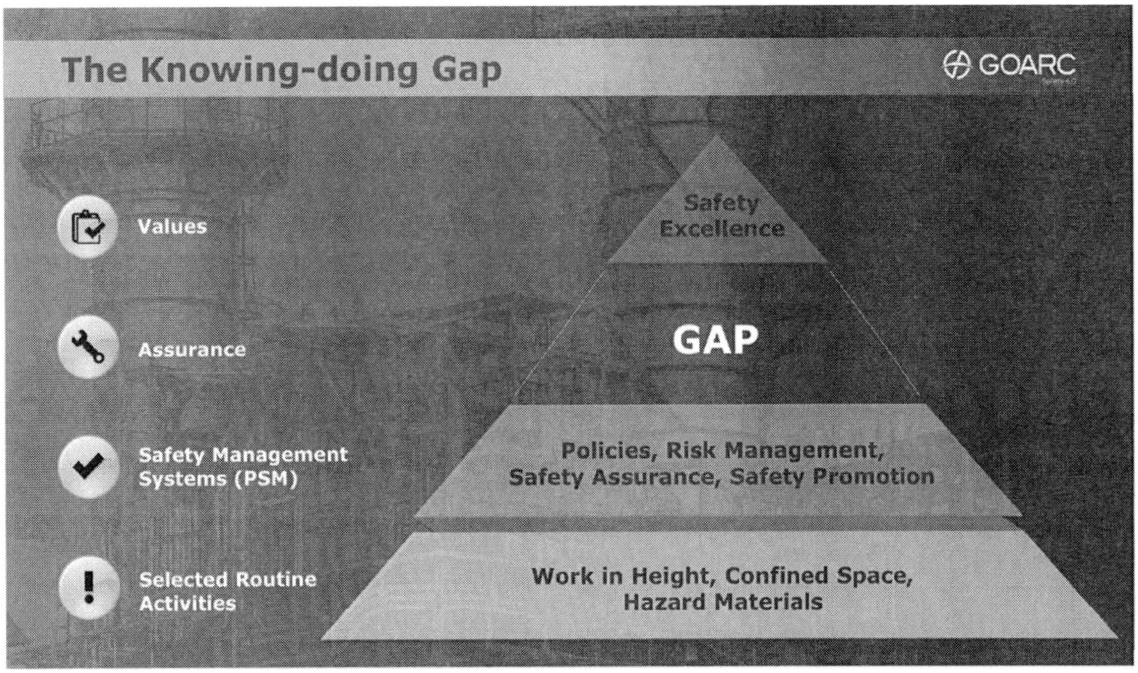
The Knowing-doing Gap
GOARC
Values
Assurance
Safety Management Systems (PSM)
Selected Routine Activities
Safety Excellence
GAP
Policies, Risk Management, Safety Assurance, Safety Promotion
Work in Height, Confined Space, Hazard Materials

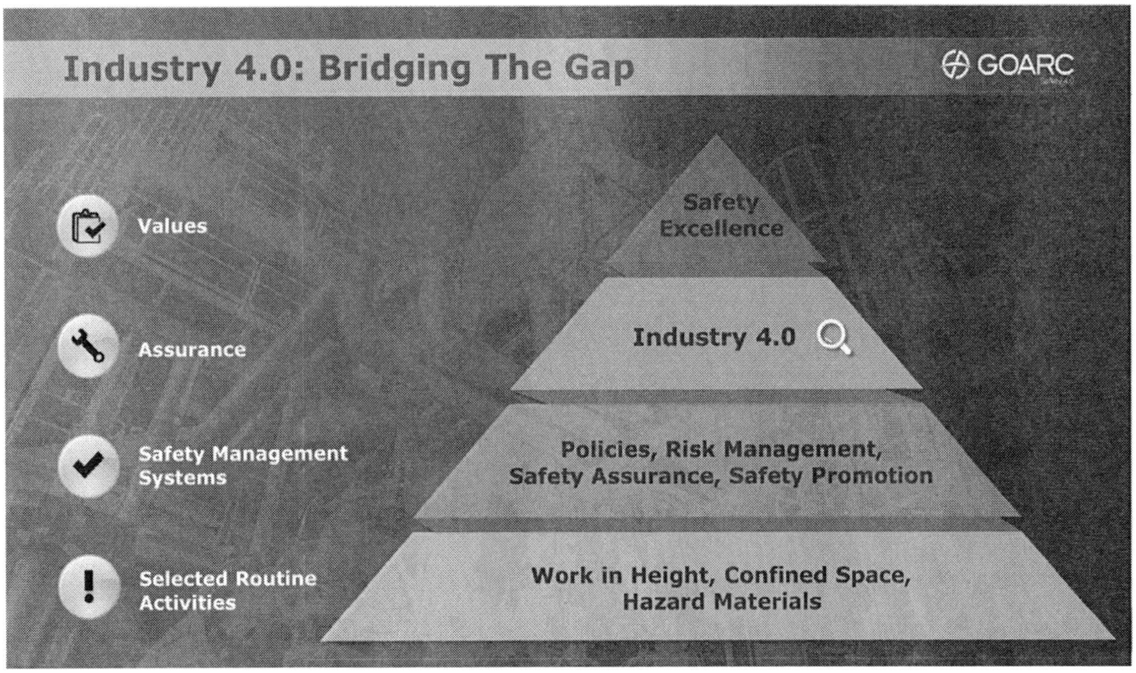
Industry 4.0: Bridging The Gap
GOARC
Values
Assurance
Safety Management Systems
Selected Routine Activities
Safety Excellence
Industry 4.0
Policies, Risk Management, Safety Assurance, Safety Promotion
Work in Height, Confined Space, Hazard Materials

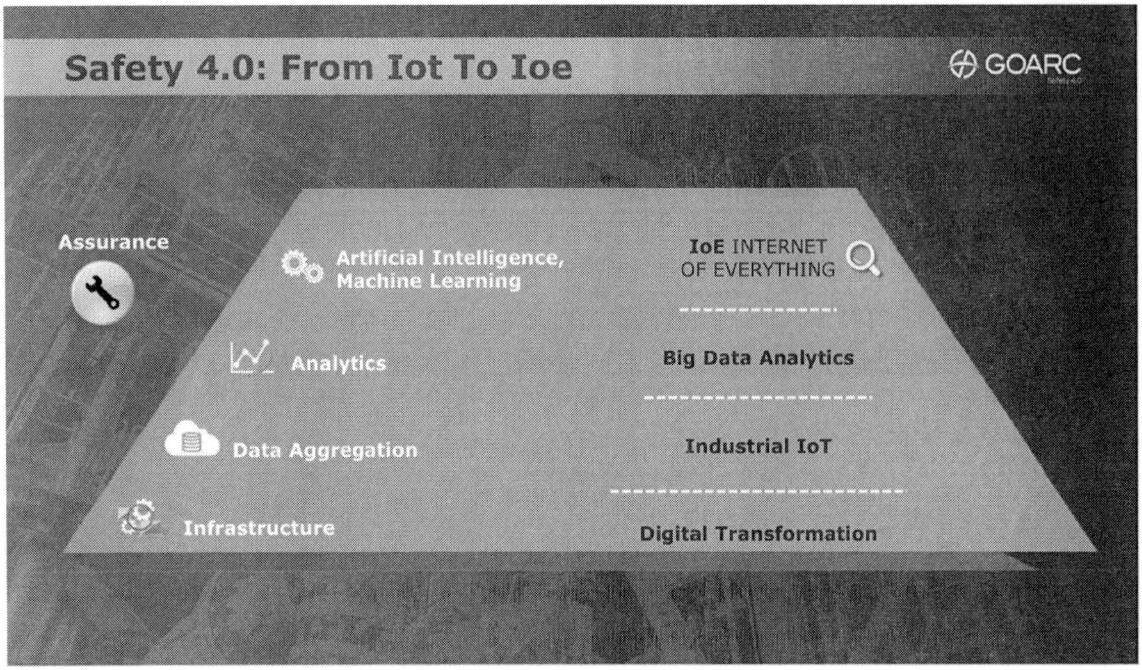
Safety 4.0: From Iot To Ioe
GOARC
Assurance
Artificial Intelligence, Machine Learning
Analytics
Data Aggregation
Infrastructure
IoE INTERNET OF EVERYTHING
Big Data Analytics
Industrial IoT
Digital Transformation

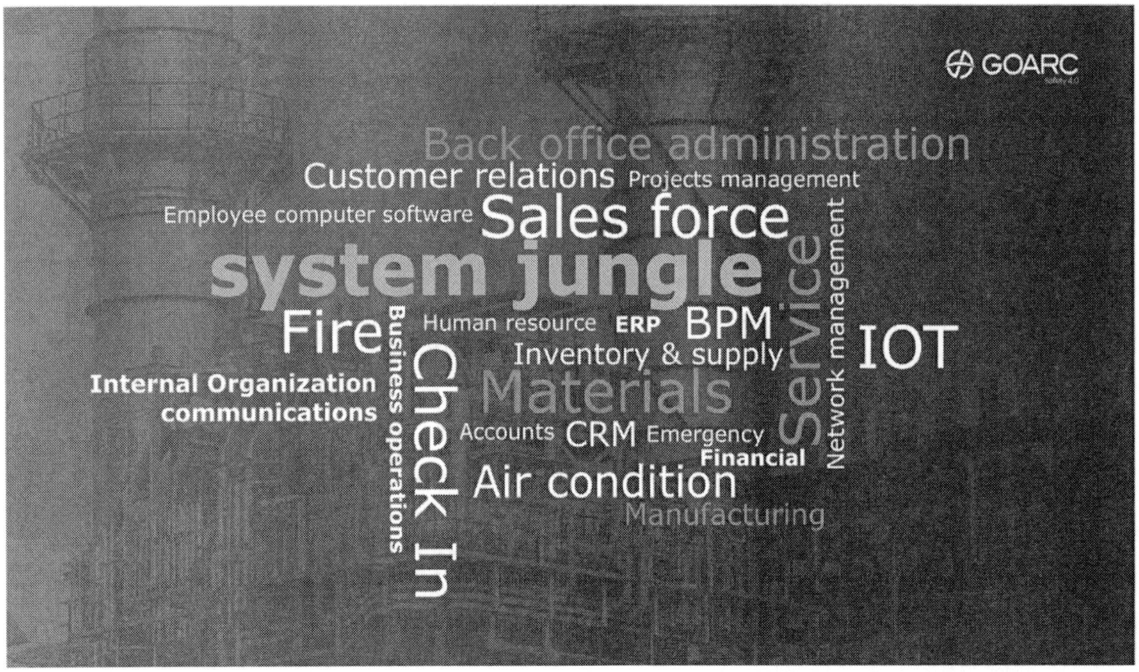
GOARC
Back office administration
Customer relations
Projects management
Employee computer software
Sales force
system jungle
Service
Network management
Fire
Business operations
Check In
Human resource
ERP
BPM
Inventory & supply
IOT
Internal Organization communications
Materials
Accounts
CRM
Emergency
Financial
Air condition
Manufacturing

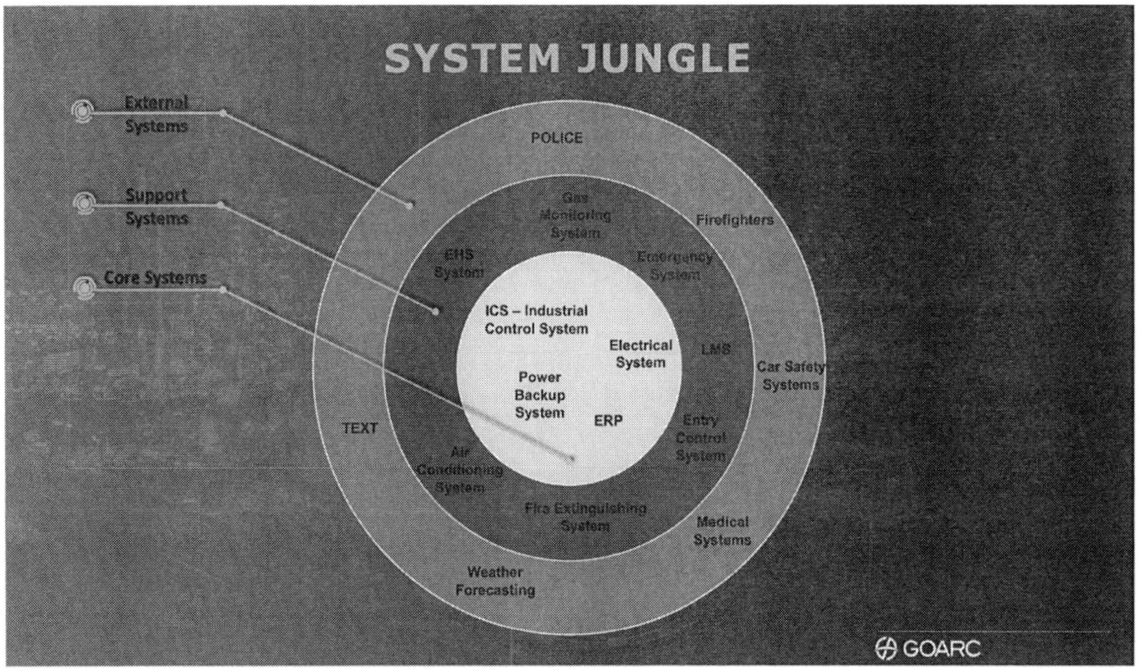
SYSTEM JUNGLE
External Systems
Support Systems
Core Systems
POLICE
Gas Monitoring System
Firefighters
EHS System
Emergency System
ICS – Industrial Control System
Electrical System
LMS
Car Safety Systems
Power Backup System
ERP
TEXT
Entry Control System
Air Conditioning System
Fire Extinguishing System
Medical Systems
Weather Forecasting
GOARC

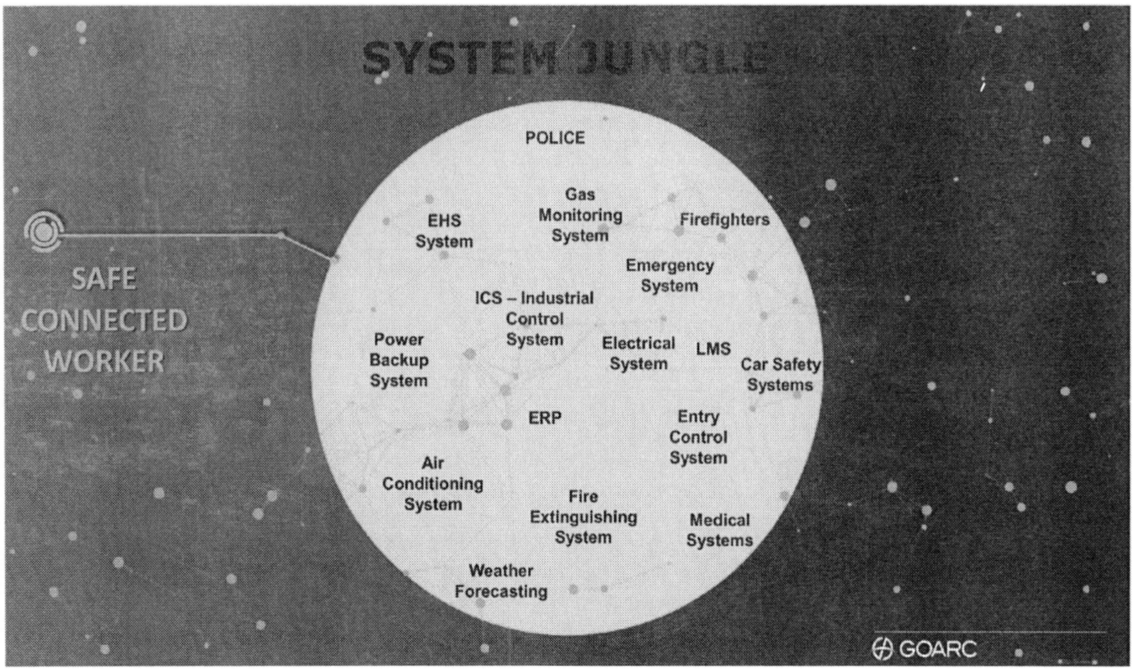
SYSTEM JUNGLE
SAFE CONNECTED WORKER
POLICE
Gas Monitoring System
EHS System
Firefighters
Emergency System
ICS – Industrial Control System
Power Backup System
Electrical System
LMS
Car Safety Systems
ERP
Entry Control System
Air Conditioning System
Fire Extinguishing System
Medical Systems
Weather Forecasting
GOARC

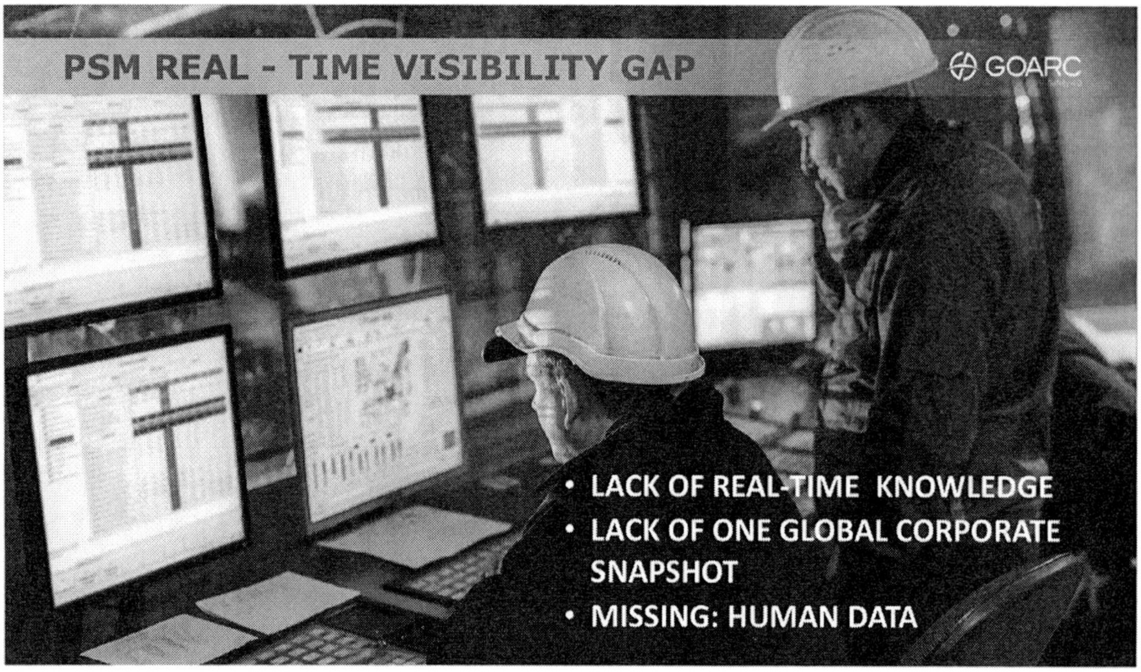
PSM REAL - TIME VISIBILITY GAP
GOARC
• LACK OF REAL-TIME KNOWLEDGE
• LACK OF ONE GLOBAL CORPORATE SNAPSHOT
• MISSING: HUMAN DATA

OPPORTUNITY - SAFE CONNECTED WORKER
GOARC
NEW STREAM OF DYNAMIC DATA:
NEW ENABLERS OF ACCIDENT PREVENTION
People
Process
Tools
Location
Materials

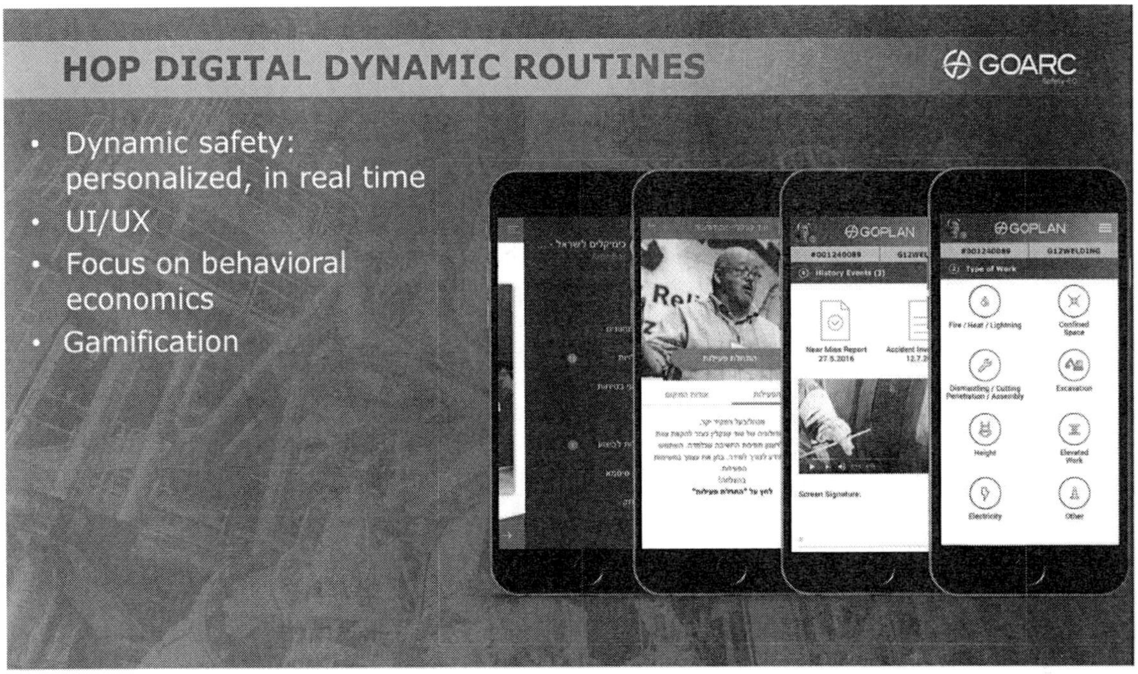
HOP DIGITAL DYNAMIC ROUTINES
GOARC
• Dynamic safety: personalized, in real time
• UI/UX
• Focus on behavioral economics
• Gamification
GOPLAN
GOPLAN

PSM 4.0
GOARC
Verify
Better Data in Real Time
Visibility

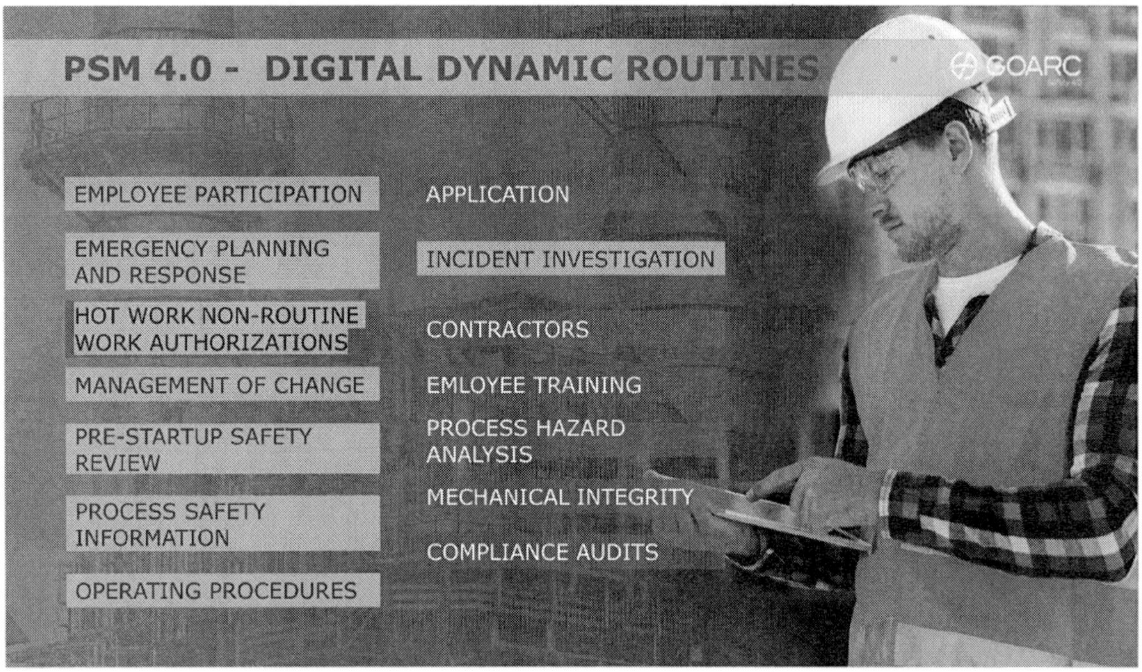
PSM 4.0 - DIGITAL DYNAMIC ROUTINES
GOARC
EMPLOYEE PARTICIPATION
EMERGENCY PLANNING AND RESPONSE
HOT WORK NON-ROUTINE WORK AUTHORIZATIONS
MANAGEMENT OF CHANGE
PRE-STARTUP SAFETY REVIEW
PROCESS SAFETY INFORMATION
OPERATING PROCEDURES
APPLICATION
INCIDENT INVESTIGATION
CONTRACTORS
EMLOYEE TRAINING
PROCESS HAZARD ANALYSIS
MECHANICAL INTEGRITY
COMPLIANCE AUDITS

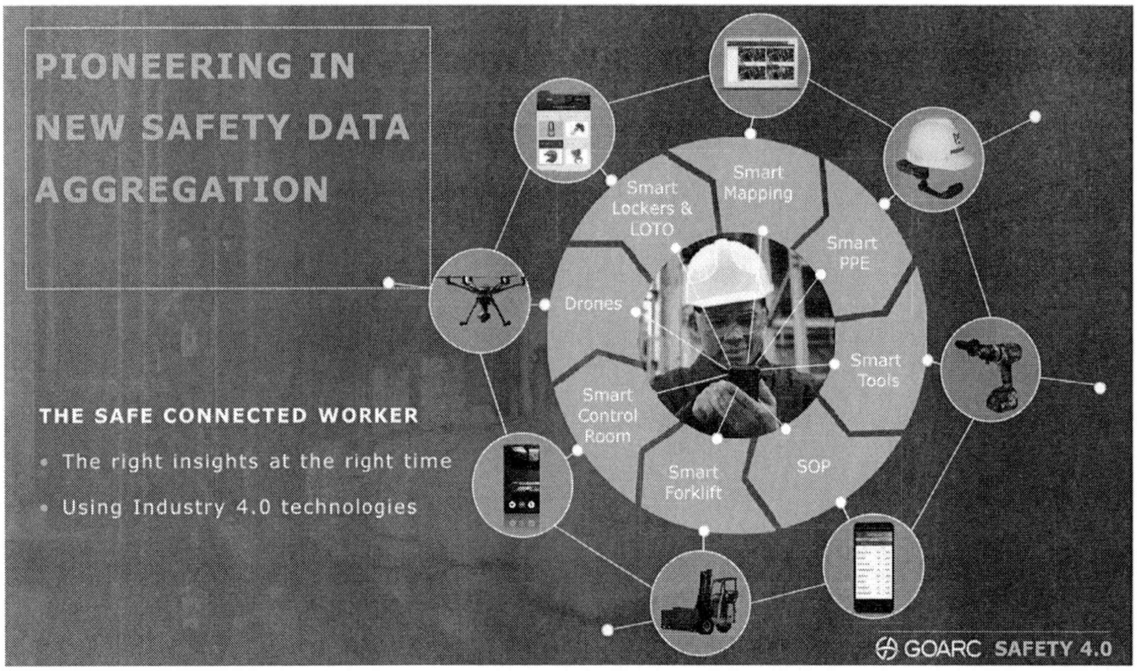
PIONEERING IN NEW SAFETY DATA AGGREGATION
THE SAFE CONNECTED WORKER
• The right insights at the right time
• Using Industry 4.0 technologies
Smart Lockers & LOTO
Smart Mapping
Smart PPE
Drones
Smart Tools
Smart Control Room
Smart Forklift
SOP
GOARC SAFETY 4.0

Converting Data Into SAFETY
GOARC
1
PREDICTION
Risk pattern recognition
2
ALERTS
Recommendations & Notifications
3
BLOCKING
People and machine work blocking
GOARC

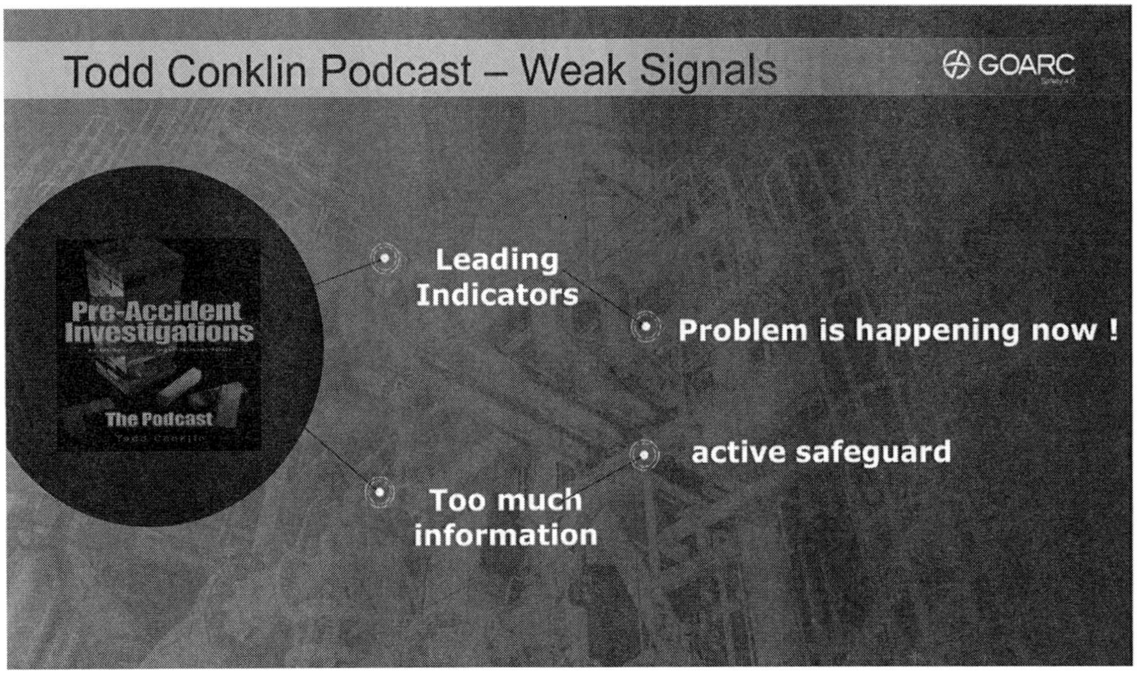
Todd Conklin Podcast – Weak Signals
GOARC
Pre-Accident Investigations
The Podcast
Todd Conklin
Leading Indicators
Problem is happening now !
active safeguard
Too much information

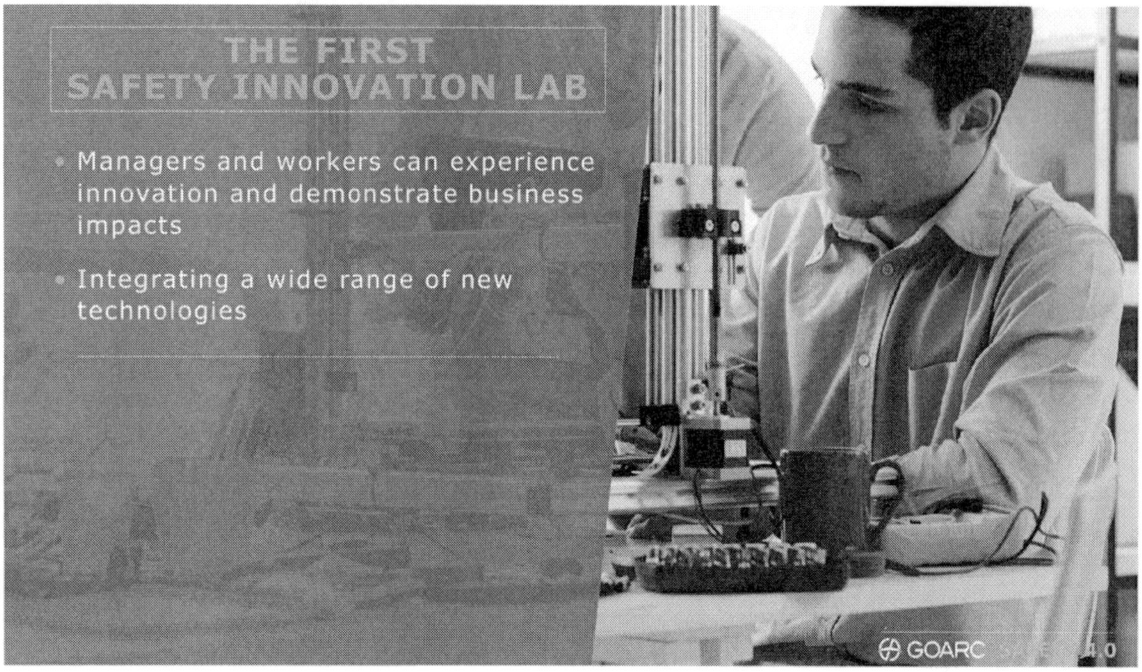
THE FIRST
SAFETY INNOVATION LAB
Managers and workers can experience innovation and demonstrate business impacts
Integrating a wide range of new technologies
GOARC

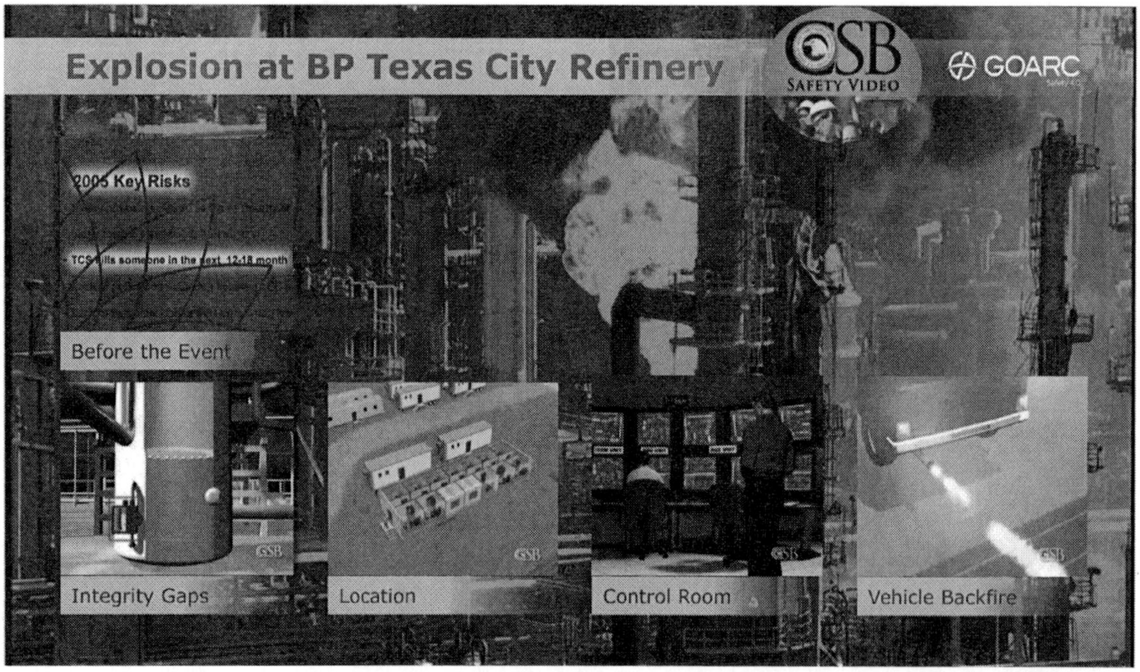
Explosion at BP Texas City Refinery
CSB
SAFETY VIDEO
GOARC
Before the Event
Integrity Gaps
Location
Control Room
Vehicle Backfire

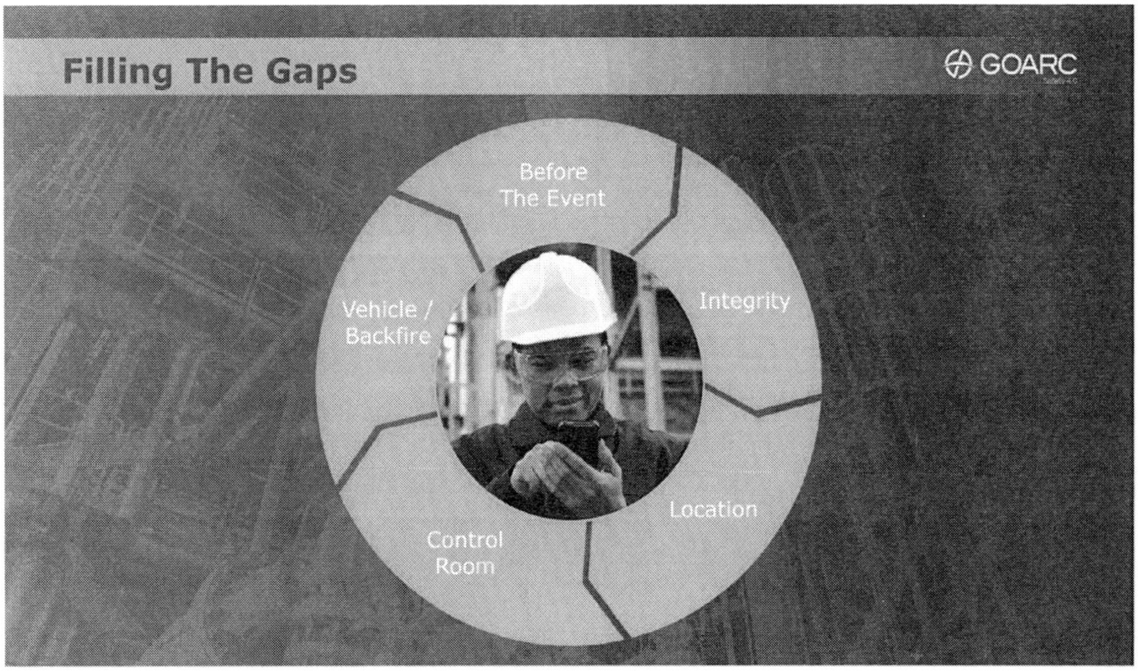
Filling The Gaps
GOARC
Before
The Event
Integrity
Location
Control
Room
Vehicle /
Backfire

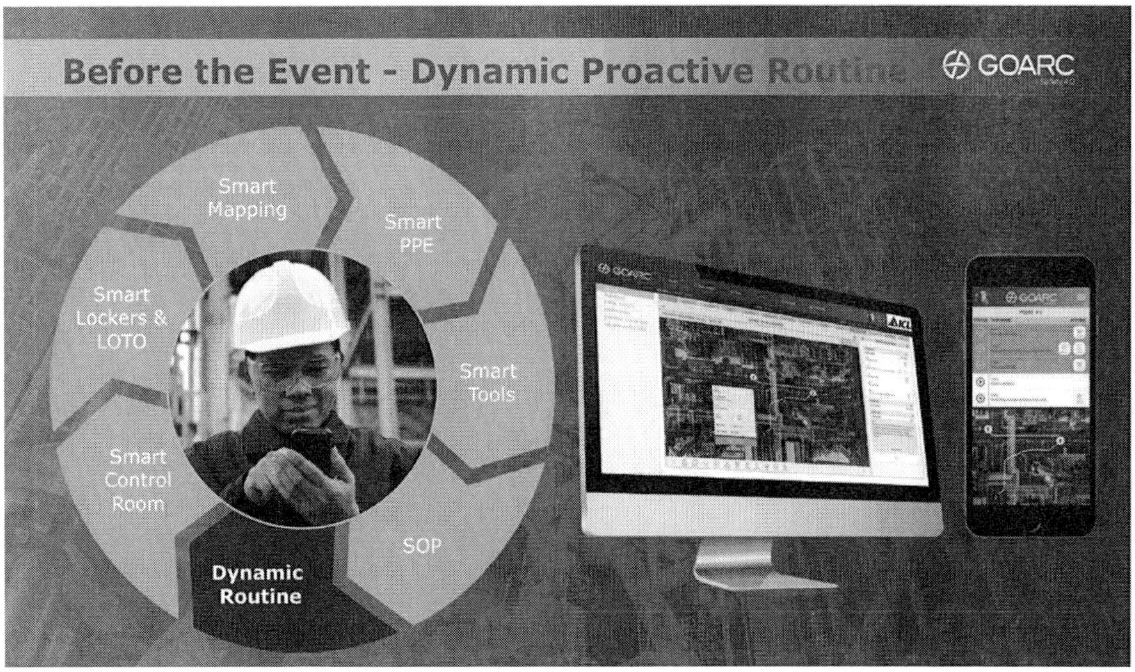
Before the Event - Dynamic Proactive Routine
GOARC
Smart
Mapping
Smart
PPE
Smart
Tools
SOP
Dynamic
Routine
Smart
Control
Room
Smart
Lockers &
LOTO

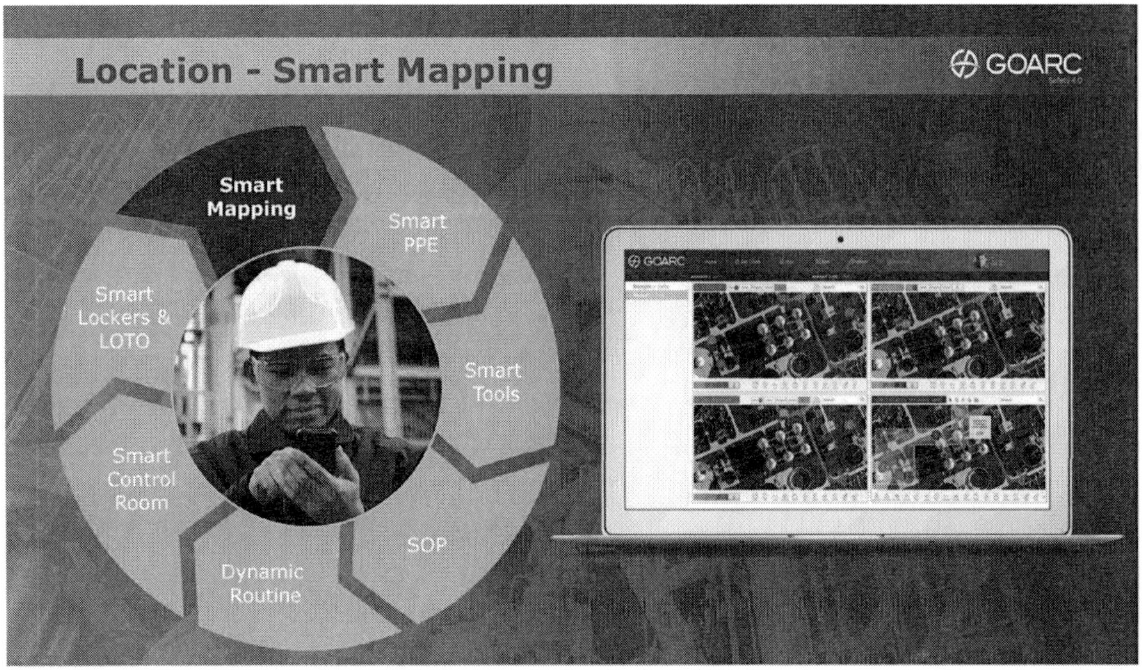
Location - Smart Mapping
GOARC
Smart Mapping
Smart PPE
Smart Tools
SOP
Dynamic Routine
Smart Control Room
Smart Lockers & LOTO
GOARC

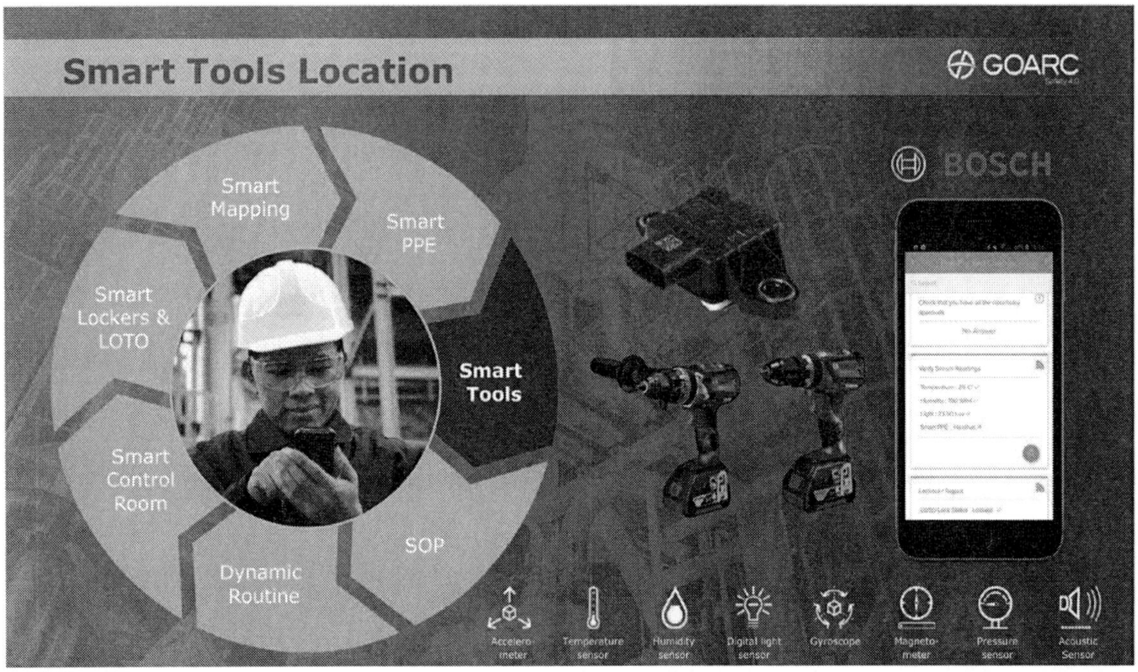
Smart Tools Location
GOARC
Smart Mapping
Smart PPE
Smart Tools
SOP
Dynamic Routine
Smart Control Room
Smart Lockers & LOTO
BOSCH
Accelero-meter
Temperature sensor
Humidity sensor
Digital light sensor
Gyroscope
Magneto-meter
Pressure sensor
Acoustic Sensor

SOP - High-Risk Processes
GOARC
Smart Mapping
Smart PPE
Smart Tools
SOP
Dynamic Routine
Smart Control Room
Smart Lockers & LOTO
GOPLAN
Emergency Information
Emergency Shower
Eye Wash
Protection Mask
Firefighting Equipment
Neutralization
PA system
Emergency Exit
Ways of Evacuation
SCANED
IS THE EYE WASH OPERATIONAL?
YES
N/A
NO
NOTES:
CLOSE

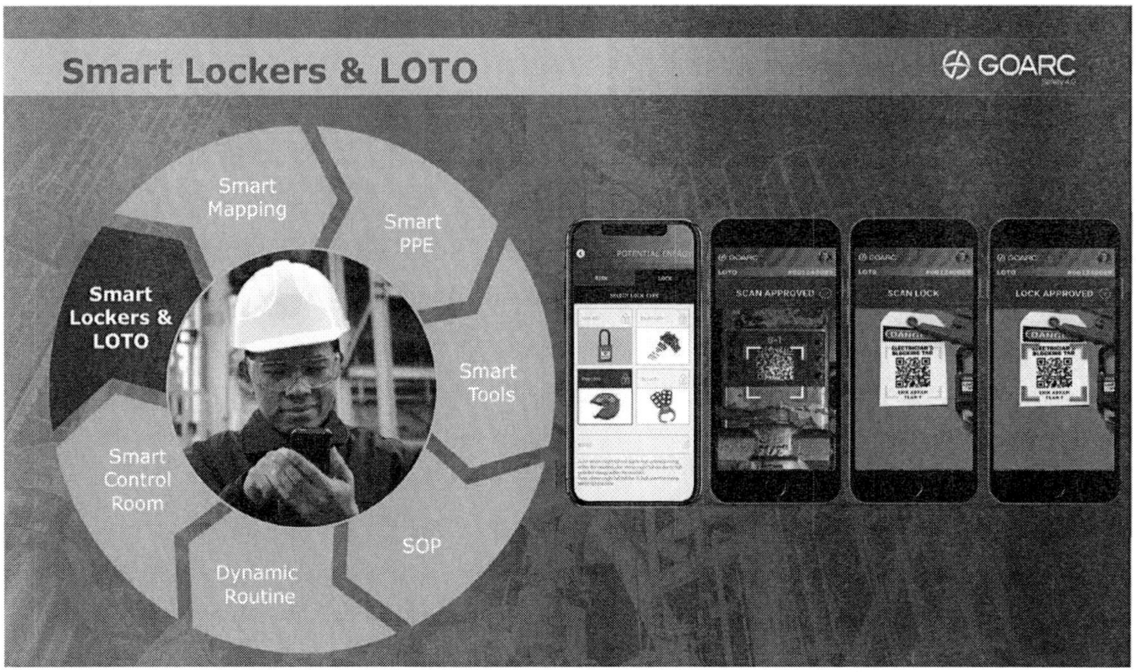
Smart Lockers & LOTO
GOARC
Smart Mapping
Smart PPE
Smart Tools
SOP
Dynamic Routine
Smart Control Room
Smart Lockers & LOTO
SCAN APPROVED
SCAN LOCK
LOCK APPROVED
DANGER

Smart Control Room
GOARC
Smart
Mapping
Smart
PPE
Smart
Lockers &
LOTO
Smart
Tools
Smart
Control
Room
SOP
Dynamic
Routine
Call
Calling...
Control room
9:41 AM

Challenges to the Process Industry Operators for leveraging Big Data for Process Safety

Panel Discussion

Marijn Bezuijen
Business Improvement Coordinator – Shell Pernis Refinery

Erik Jan van den Heuvel
Continuous Improvement Center of Excellence - Dow

Nicholas Sands
Senior Manufacturing Technology Fellow - DuPont

Nicolas Ranchet
Refinery 4.0 – Technical area business owner - Total

Moderator: Richard Pawulski, PE, P Eng
Vice President, Center for Technical Expertise - Air Liquide USA

1

Richard Pawulski

Vice President, Center for Technical Expertise - Air Liquide USA

Responsibilities
Manage Technology Subject Matter Experts
Support North American Plant Operations

Link to Process Safety and Big Data
Supporting Air Liquide's Big Data initiative – Smart Innovative Operations (SIO)
Deployment of Air Liquide's Industrial Management System (IMS)
Managed corporate functions responsible for development and deployment of Process Safety Rules

Affiliations
Compressed Gas Association – Air Liquide company representative
Center for Chemical Process Safety - Air Liquide company representative

email: richard.pawulski@airliquide.com

2

Marijn Bezuijen

Business Improvement Coordinator - Shell Pernis Refinery

Responsibilities

Digital Lead for the Pernis Refinery
Global Manufacturing Business Product Owner for Connected Maintenance Execution

Link to Process Safety and Big Data

Sponsor, Project Lead or Product Owner for a number of Shell's strategic Digitalisation initiatives

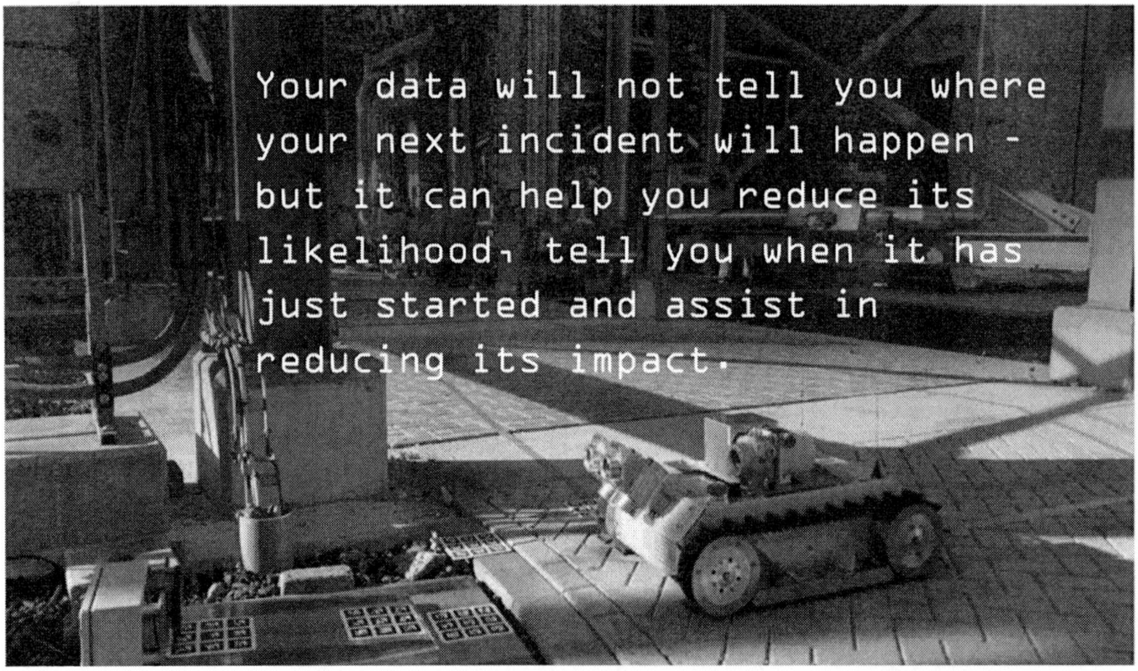

Digital innovation in Process Safety drives other business outcomes

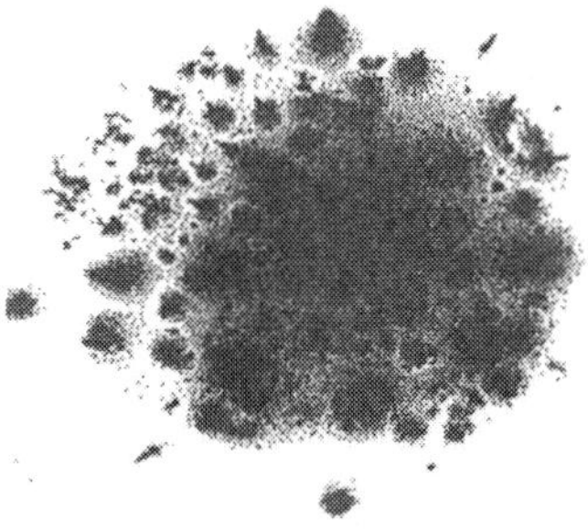

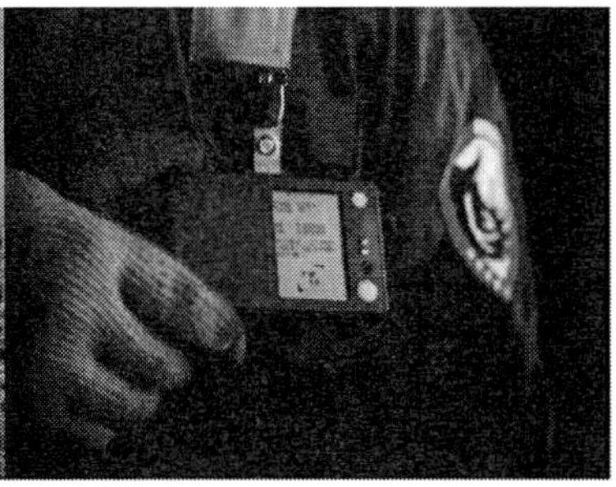

Predictive maintenance

Started for process safety, rolled out widely for reliability

Automatic leak detection

Started for process safety, replicated for reputation

People localization

Started for process safety, invested in for execution efficiency

---

Erik Jan van den Heuvel

Continuous Improvement Center of Excellence - Dow

Current Responsibilities

Industry 4.0 site leader for Terneuzen (NL)
Industry 4.0 Analytics EMEAI & Pacific
Global IT Leader for Manufacturing Labs
AI (ML / DL) focal point in Operations

email: evdheuvel@dow.com

6

Industry 4.0 in Dow

Global initiative ("Digital Dow") → in Operations: "Manufacturing 4.0"

Platforms: *Analytics, Mobile, Robotics, Digital Thread, Sensors*

Sensors	Visibility, CUI, manual valve position
Analytics	AI (modelling 'normal') & Predictive maintenance
Mobile	Remote expertise (VR)

Some (process) safety related initiatives

- Train decompression events → look @ process data
- Predictive Maintenance (Asset Health)
- Elimination of CSEs, elevated work, inspections in pipes/ vessels
    - Robots, drones, rope-access
    - CSE inspections by humans require an exception case in Dow!
- Remediation → Combining P&IDs & scanning for AR: avoid exposure

7

Industry 4.0 in Dow

Localization → (gas) sensors & analytics → safety

- Bonus: wrench time optimization

Using LoRa

LoRa Technology is a long range, low power wireless technology used in IoT

- Tracking equipment and people?
- Manual valve position
- Corrosion
- Wearables?

Digital procedures

- 'Tuned' based on user experience → Right information / Right place / Right time
- More realistic / specific: adjusted to actual workplace via AR

Electronic Safe Work permits → including all disciplines → no omissions

EH&S: Analytics dashboard with Leading indicators → trends in (severe) near misses

- Focus EMI

Pitfalls: the human element, culture, technology-for-the-sake-of-technology

8

Nicholas Sands

Senior Manufacturing Technology Fellow at DuPont

30 years of Automation and Process Safety experience

Roles

- Global Alarm Management Leader for DuPont
- Currently working in Luxembourg
- Currently living in Dallas, Texas
- Certified Automation Professional
- International Society of Automation (ISA) Fellow
- Process Automation Hall of Fame
- Lead for many Automation Best Practices in DuPont

ISA/IEC STANDARDS

- Co-Chair of ISA18 on Alarm Management
- Co-Director of ISA84 on Safety Instrumented Systems
- Co-Director of ISA101 on Human Machine Interface
- Secretary of IEC62682 committee on AM, lead editor
- Member of other ISA standards: ISA105, ISA108...
- Editor of Guide to Automation Body of Knowledge 3rd edition

Nicholas.P.Sands@dupont.com

9

Enterprise Alarm Management – A Starting Point

Enterprise wide alarm and event data collection and reporting:

- Since 2010
- At the peak, ~200 plants
- Many different systems

Lessons learned:

- Constant support for changes (software, people, systems)
- Verify accuracy of data

Data often includes:

- Operator actions
- Automation events

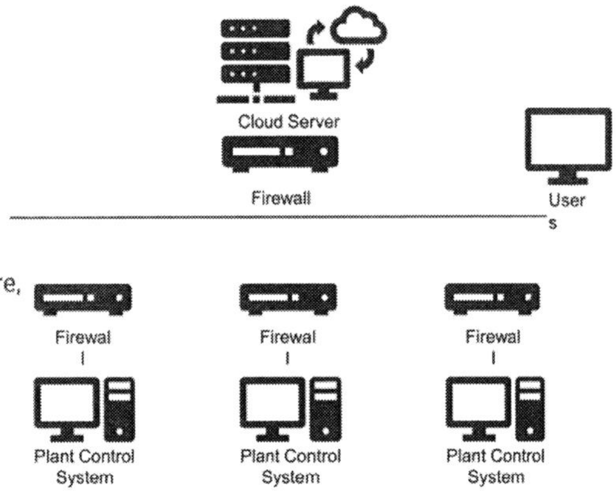

10

The Next Steps for Process Safety

Sequence Verification

Comparison of automation events against expected sequence and timing:
- Early detection of abnormal conditions
- Identify opportunities for improved automation

Examples:
- Burner lightoff
- Reactor start-up

Operator Effectiveness

Tracking of operator actions during normal and abnormal events:
- Detect need for improved training
- Identify opportunities for improved automation and HMI

Examples:
- Correct and incorrect actions
- Human Machine interface actions (screen changes…)

Trends

Develop KPIs for sequence verification, operator effectiveness, and more.

Trend KPIs through changes in organization and process technology.

Examples:
- Operator training after personnel changes
- Increased operator interventions after process technology change

Nicolas Ranchet
TOTAL Refining & Chemicals

R&D, Center in India for Digital Innovation (CIDI), Refinery 4.0 program
Focus Area Manager, Safety, Energy Efficiency and Availability

Link to Process safety and Big Data
Lead the development of an online failure mode effects analysis advisor and an alarms management application.

email: nicolas.ranchet@total.com

TOTAL

TOTAL Refining & Chemicals
Refinery 4.0 program

- Safety
- Competitiveness
- Environmental footprint
- Agility

Inventing and developing Digital Innovations, benefiting from experiences across industries

- Artificial Intelligence, Data Science
- Internet of Things, Edge computing
- Automation, Interoperability
- Optimization, decentralised decisions

To transform our way to operate, supervise, inspect, maintain, optimize the refinery.

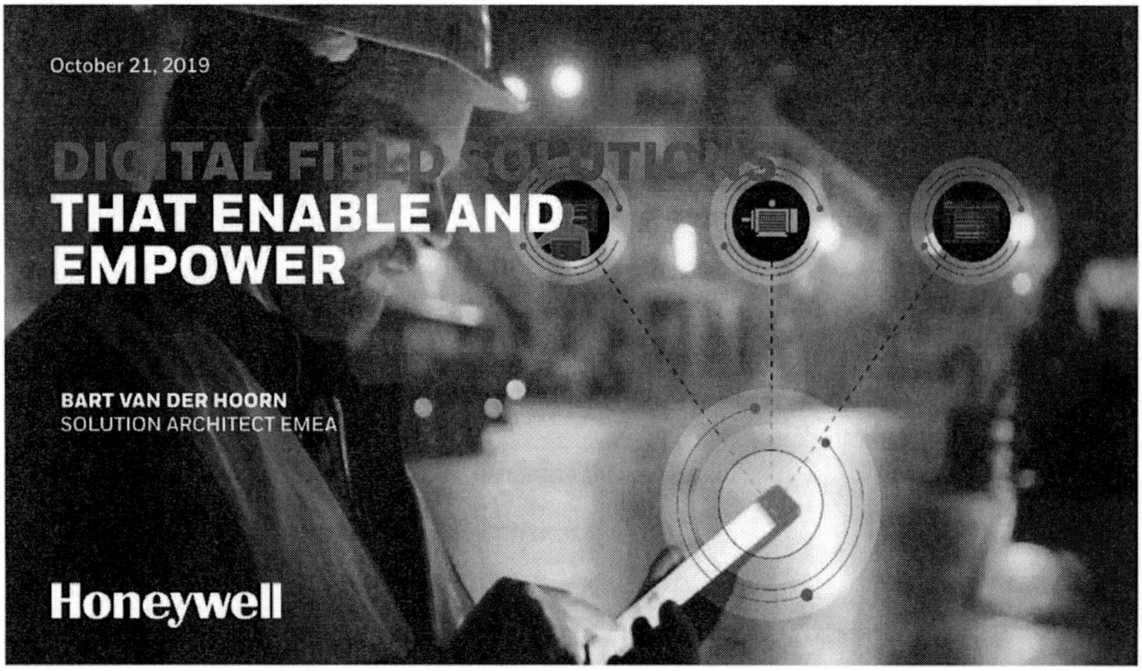

INTRO

Bart van der Hoorn
Honeywell the Netherlands (Schiphol/Amsterdam)

2007 - Intern/vacation work/graduation

2010 - Junior Engineer Training Program (JETPro)

2011 - (sr.) Design Engineer

2014 - HMI/operator workspace specialist - COE Operator Effectiveness

Global champion for HMI's that deliver increased operator effectiveness over a greater scope of responsibility, faster and safer responses, while reducing fatigue and increasing situational awareness.

2018 - Solution Architect for Connected People

Responsible for the strategy, design and implementation of Honeywell's competency, collaboration and compliance outcome programs that drive safety, production and cost reduction KPIs.

1

ABNORMAL SITUATION MANAGEMENT (ASM)

Founded in
1994
to address customer concerns

The high cost of incidents | Unplanned shutdowns | Fires | Explosions | Emission releases | Accidents

Unique Consortium Member Model

- **User Members** – Industry needs
- **University Members** – Technology know how
- **Associate Members** – Consultants & Experts

ASM

Research and Development Consortium by a group of leading companies and universities

ASM OBJECTIVES

Understanding Abnormal Situations
This area focuses on measuring, reporting, analyzing, and communicating the causes and effects associated with abnormal situations.

Organizational
This area focuses on management practices that influence the organizational culture, work processes, staff roles and responsibilities, and valued behaviors as they relate to abnormal situations.

Knowledge
This area focuses on the development and maintenance of a competent work force through training and the creation of a continuous learning environment so that personnel can effectively respond and cope with abnormal situations.

Communications
Focuses on daily communication and situational dialog among plant personnel and explores opportunities to use information technology that improves site-wide coordination in all situations.

ASM CATEGORIES

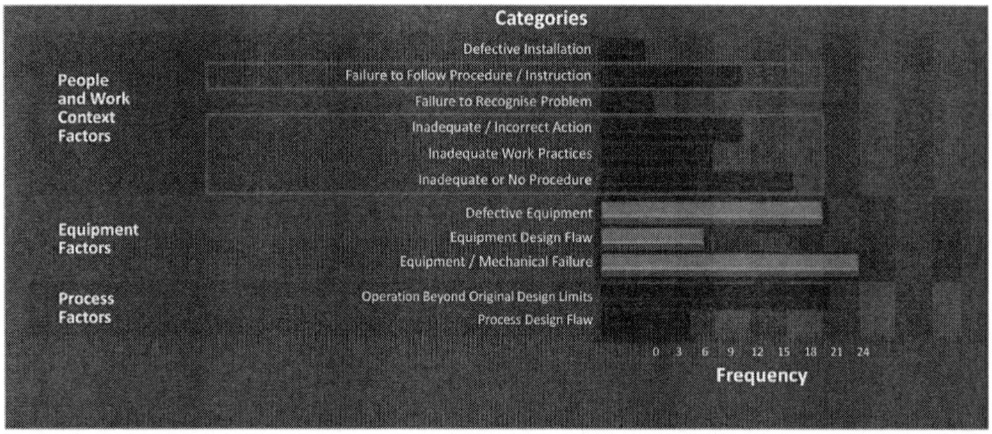

57% of errors are procedure related

 4

OIL & GAS CUSTOMER WORKFLOW DIGITIZATION

Honeywell 5

A DAY IN THE LIFE OF A FIELD WORKER

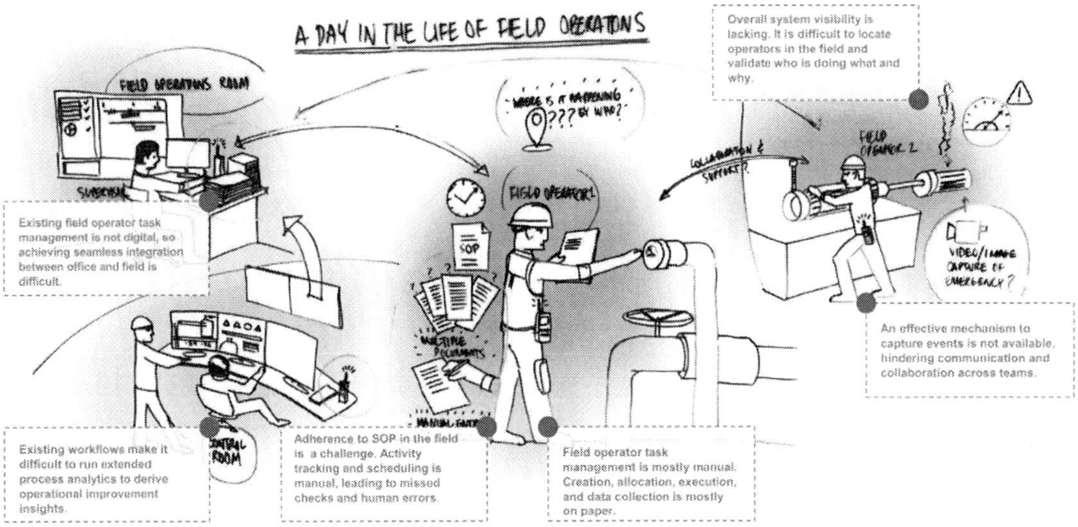

 6

THE CHALLENGES

Manual Data Gathering

Most activity tracking/scheduling is manual and paper-based, leading to missed checks and errors.

Limited End-to-End Visibility

No visibility during inspection process. Field technician whereabouts and status are unknown.

Downtime and Inefficiencies

Routes, workflows, and processes are not optimized for peak performance, leading to downtime, inefficiencies, and time wasted between tasks.

Operator rounds data is not stored or used for analytics!

 7

FUNCTIONAL SCOPE

Project Size

24x7 production facility

Safety & Reliability main stakeholder

Operator Driven Design

Process / Functions

- Operations
- Maintenance
  - Instrumentation
  - Electrical
  - Mechanical
- Safety
- Fire
- Security

Field Activities

- Structured Field Rounds
- Operation Checklist
- Ad-hoc Field Tasks
- LLF Rounds
- Maintenance Checklist
- Equipment Service Sheets
- Thickness Survey
- Fire Equipment Survey
- Safety Observation Program
- Self Verification Audit

SIGNIFICANCE OF OPERATOR FIELD ROUNDS PROJECT

Robust Data Collection

- Automate data collection activity of field rounds by Operations and Maintenance to help improve accurate and timely data collection
- Help in capturing the data of Non-DCS tags (such tags will be a P&ID tag) in a central database for monitoring equipment condition and process monitoring

Better Planning

- Make process of scheduling and planning of field rounds by Operations and Maintenance Personnel more efficient.
- Increase adherence to the field rounds by timely execution at desired frequencies

Enhanced Visibility for Quicker Actions

- Alert field personnel in real time when input data is beyond the defined limits. Highlight consequences of deviations & provide key actions to be taken to normalize the parameter online
- Enable speedy data review by Operations / Maintenance / CTS engineers
- Visibility of field deviations in shift logbooks for further review and escalation
- Availability of field images for better decision making

Integration with Maintenance Applications

- **Allow work request (in the form of PM Notifications) to be triggered directly from field.**

CHALLENGES IN FINDING THE RIGHT SOLUTION

Solution should have:

- Capability to capture various field activities carried out by Operations, Maintenance, HSE and any other function in manufacturing
- A robust planning and scheduling tool to meet the dynamic need of plant operations
- Ease of building templates, workflows and integrations with external systems
- Scalability to multiple sites and plants
- Futuristic, cloud based architecture
- Native, cross platform and device agnostic mobile app

Device and Communication:

- Device should be Intrinsically Safe for use in the Hydrocarbon industry
- Future proof device that runs on latest Android / iOS / Windows OS
- Data network across plants to enable real time communications

 10

FUNCTIONAL ARCHITECTURE

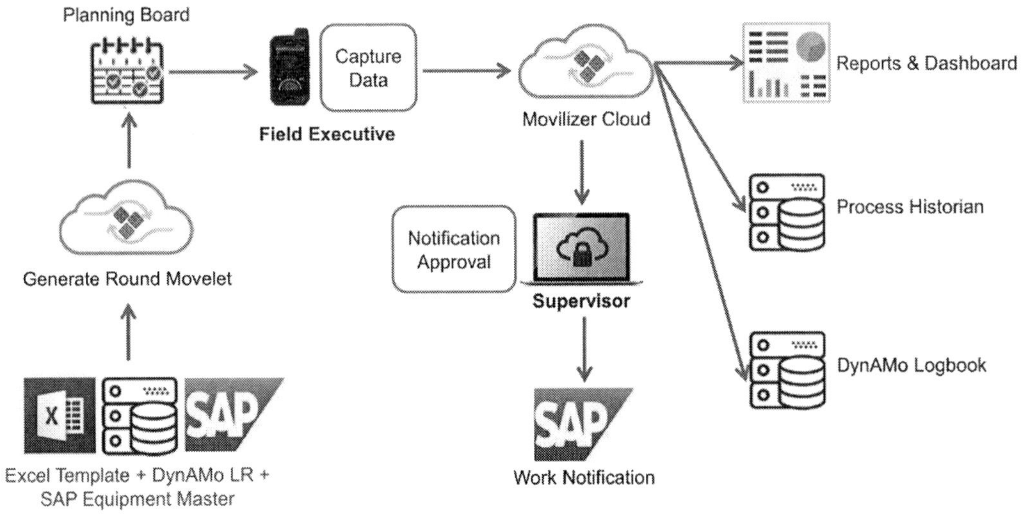

 11

HONEYWELL FORGE INSPECTION ROUNDS

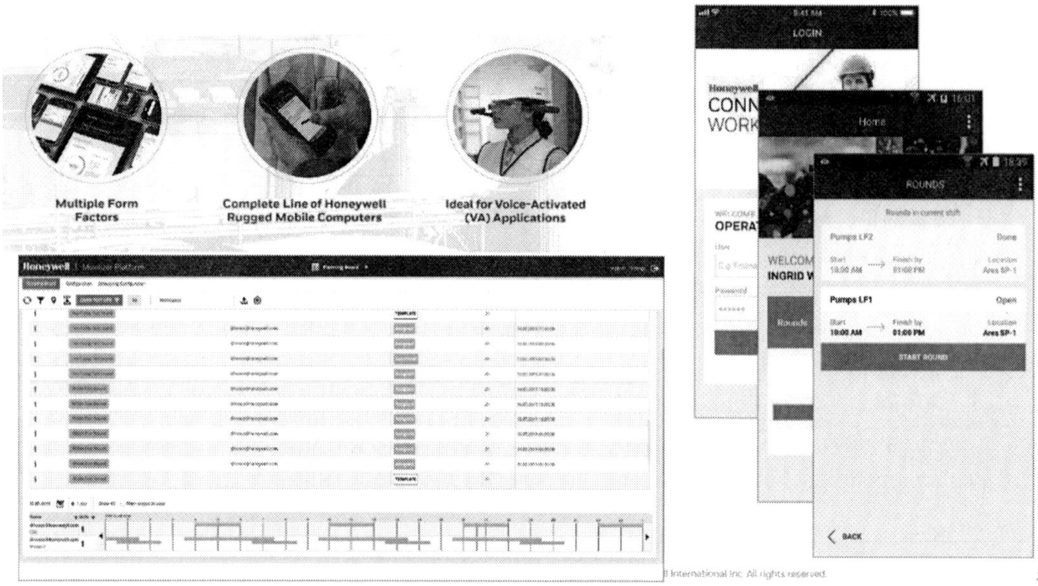

 12

RESULTS - FUNCTIONAL

- Automated data collection activity of field rounds and thereby improve data accuracy and timely data collection.
- Alerting the Field Executives in real time when input data is beyond the operating limits. Highlight consequences of deviations & provide key actions to be taken to normalize the parameter online.
- Enable Operation, Technology and Maintenance team to monitor equipment performance and to do proactive process monitoring.
- Helps in rationalization of rounds with historical data analysis.
- Facilitate Field Executives to trigger Work Notifications directly from field.
- Support and help effective field audits that need to be carried out by Safety, O&M etc.
- Helps in building central limit repository by providing integration with DynAMo LR.

 13

RESULTS - TECHNICAL

- Enterprise Private Cloud Mobile application is aligned with customer's future plan
- Single scalable infrastructure with a common solution and architecture helps in business process standardization across different business domains
- Single instance (centralized) installation has significant savings in hardware & maintenance cost
- App works across all mobile OS (Android, iOS, Windows) & platforms (phones, tablets) and hence greater flexibility in introducing new workflow functions as required
- The cloud platform has capabilities to interface to most of the existing enterprise platforms through a single instance which offers greater flexibility of data availability

FUTURE PLAN

- Potential integration with various tools
  - vibration probes,
  - temperature gun etc.
- Interface with Central Document System to provide on-demand documents to field executives.
- Build analytics on equipment reliability using the data from
  - Movilizer,
  - SAP and
  - Historian.
- Secure video streaming from field.
- GPS tracking and geofencing
- Positive NFC (Near-Field Communication) validation of assets

PROCESS SAFETY SUITE WORKFLOW INTEGRATION

Honeywell

16

USE CASE: COLLECT & USE "AS FOUND" TEST DATA

<u>Full</u> Safety Lifecycle
Compliant with ISA-84 / IEC 61511

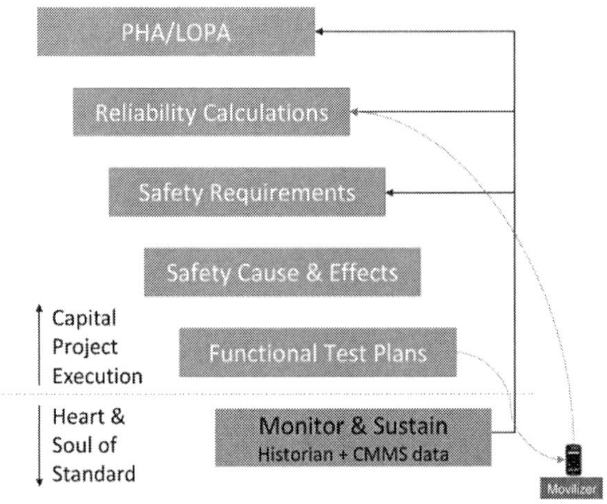

AUTOMATING THE DATA-FLOW IN THE PROCESS SAFETY LIFECYCLE

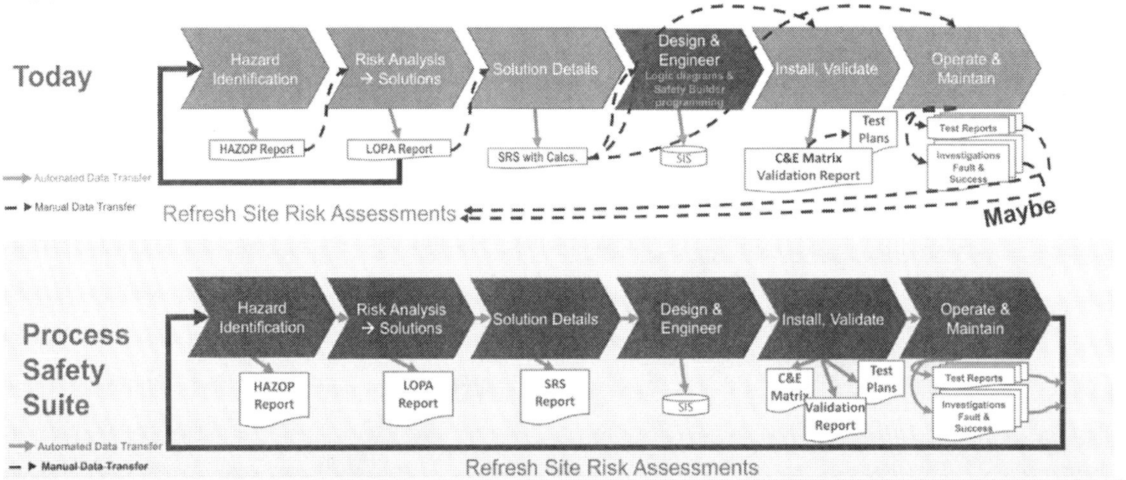

Data all in a single "model". No manual transfer from step to step.

CONNECTED INSTRUMENT INSPECTION SYSTEM

Past flow

- CMMS indicates Inspection is due
- Inspection procedure sent to Inspector
- Field inspection completed.
- Today, paper goes into file drawer

New flow

- Honeywell Forge Inspection Rounds on Honeywell CT60 ATEX zone 2 device for IPL Testing in plant
- As Found/As Left data collected via App on device and send back to PSW
- Validate Maintenance assumptions

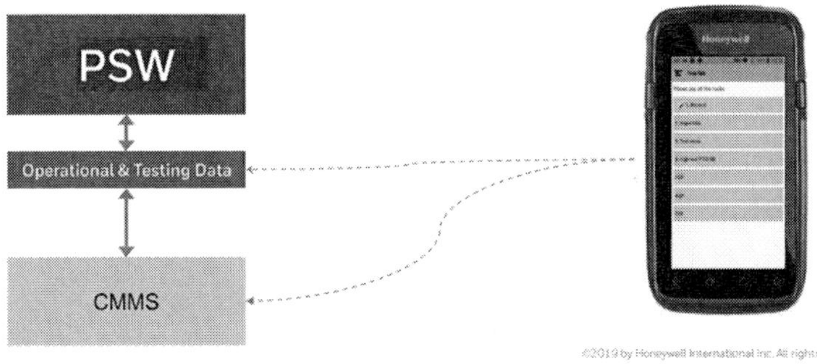

19

IPL EXAMPLE (1)

Task	Step	Staff	✓	Work Method Steps
IPL Test Plan - Work Methods for IPL: SIF-HIPPS-01				
1				PRE-TEST TASK - Please review the following steps in order to prepare for functionally testing the IPL.
	1		☐	Obtain the appropriate permits for the testing and provide the proper notifications to operating personnel regarding the testing, any special requirements, and subsequent alarms that may be generated.
	2	OP/IT	☐	Verify that any special safety procedures associated with testing or the actions required during testing have been identified and are being followed
	3	IT	☐	Verify any required personal protection equipment (PPE) required for the testing is available and in good condition.
	4	IT	☐	Verify that the test equipment to be used is in safe and good operating condition and has been calibrated within the past 12 months.
	5	IT	☐	Visually verify field installation of all components (switch, instrument and electrical installations, process equipment, valves, tags, etc.) associated with this test are in sound condition and accessible. If necessary, follow proper procedures to obtain access. Record device condition in the "As Found" line on page 1.
	6	IT/OP	☐	If applicable, refer to Cause and Effect drawing to ensure that all causes and effects are known.
2				INSPECTION -
	1	IT	☐	Visually verify installation of all components (devices, electrical connections, physical support, tags, etc.) associated with this test are in sound condition and accessible.
	2	IT	☐	If applicable, remove insulation and inspect condition. Ensure heat tracing is properly installed.
	3	IT	☐	Inspect all process connections and motive force connections to ensure they do not show any damage, leaks, or degradation.
	4	IT	☐	Review all isolation valves, drain / vents and manifold valves to confirm pressure transmitter is properly aligned to read process conditions. If device is not properly aligned record as a failure.
	5	IT	☐	Record device condition in the "As Found" line in the results section of this document and write work orders to fix any damaged devices or equipment.
3				TEST SETUP -
	1	IT / OP	☐	Coordinate with operations to install bypasses (software, jumpers, mechanical stops etc..) if required to prevent process upset upon trip.
	2	IT	☐	Close the root valve at the process connection for PT-0100 . Note that there may be more than two bridle isolation valves. Verify the installation before proceeding.
	3	IT	☐	Close the isolation valve on the manifold for PT-0100 and open the vent valve.
	4	IT	☐	Connect a pressure calibration system to PT-0100 and close the vent valve.
	5	IT	☐	Connect a HART Communicator to PT-0100
	6	IT	☐	Connect a reference multi-meter to PT-0100 to read the output current.

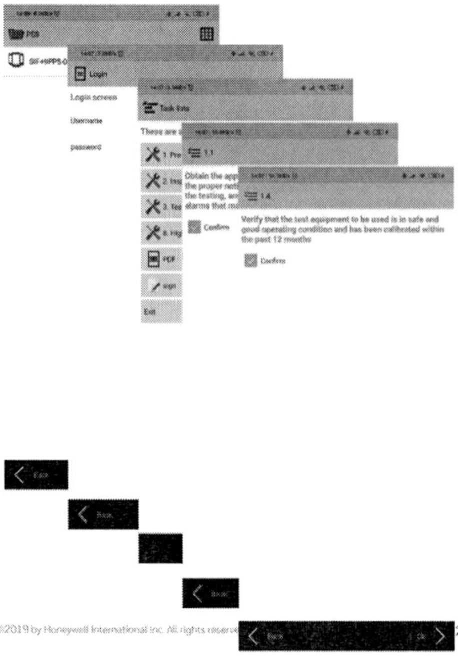

IPL EXAMPLE (2)

Task	Step	Staff	✓	Work Method Steps
8				HIGH TEST -
	1	IT	☐	Confirm upon initial setup that the security jumper/switch is enabled and no transmitter fault alarms or warnings are present. If device is abnormal record as a failure.
	2	IT / OP	☐	Apply 0 PSIG and verify the output reads 4mA and process graphic reads correctly. If it does not, mark the device as a Fail and continue the test.
	3	IT / OP	☐	Reset any trips and alarms that are present for PT-0100 . Verify that all final elements are in their normal operating condition.
	4	IT / OP	☐	Slowly increase the pressure on PT-0100 above high trip point of 550 and confirm the trip state on the process graphic.
	5	IT / OP	☐	Verify and acknowledge the alarm indication
	6	IT / OP	☐	Record the observed trip point and compare it to the intended trip point. If the trip point is not within the tolerable range, mark the device as a Fail and continue the test. PT-0100 Observed High Trip Point: ___________ Upper Test Limit: 650 Lower Test Limit: 0

Remember:

- Set up is not difficult
- Collect As Found Data
- Must USE the DATA

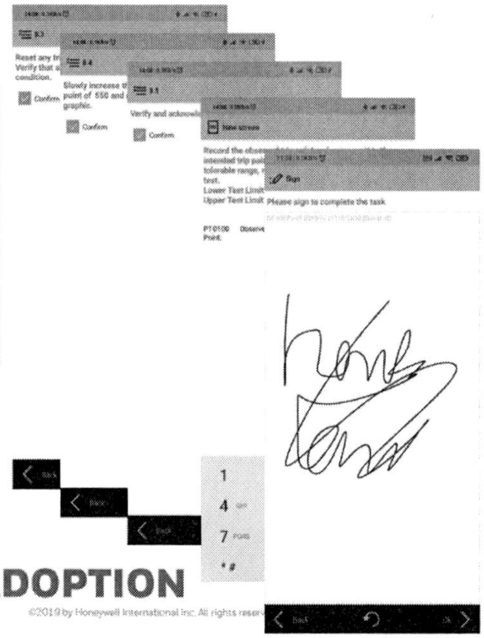

EASY USE LEADS TO FIELD ADOPTION

THANK YOU

Honeywell 22

Industry 4.0 for Process Safety

2019 European Conference
Process Safety and Big Data
Center for Chemical Process Safety

Jim Petrusich

July 2019

Explosion at China Gas Factory

-15 People killed, and 15 injured

-Air separation unit of a gas factory

March 21 2019

Pesticide Plant Explosion China

-78 People killed, and more than 600 injured

-Chemical plant in Xiangshui located in Jiangsu province

November 2018

China Chemical Park Explosion

-24 People killed, and 21 injured in Zhangjiakou (host city of 2022 Winter Olympics)

-Leaked chloroethene (vinyl chloride) set off a chain of 38 truck explosions.

HOUSTON CHRONICLE

Explosion, fire at [redacted] Baytown plant injures 37

A fire at an [redacted] plant in Baytown Wednesday morning injured 37 people and sent a plume of smoke over a Houston-area chemical facility for the fourth time since April.

It was the second fire this year at an [redacted] facility in Baytown. The company's operations in the east Harris County city have a history of environmental violations stretching back to 2013.

A Houston Chronicle investigation from 2016 found that a major chemical incident occurs once every six weeks in the greater Houston area. Advocates said Wednesday's fire again demonstrated the need for stronger enforcement of safety and environmental rules.

A major chemical incident occurs once every six weeks in the greater Houston area.

Deer Park Fire -March 2019

Learnings from Texas City

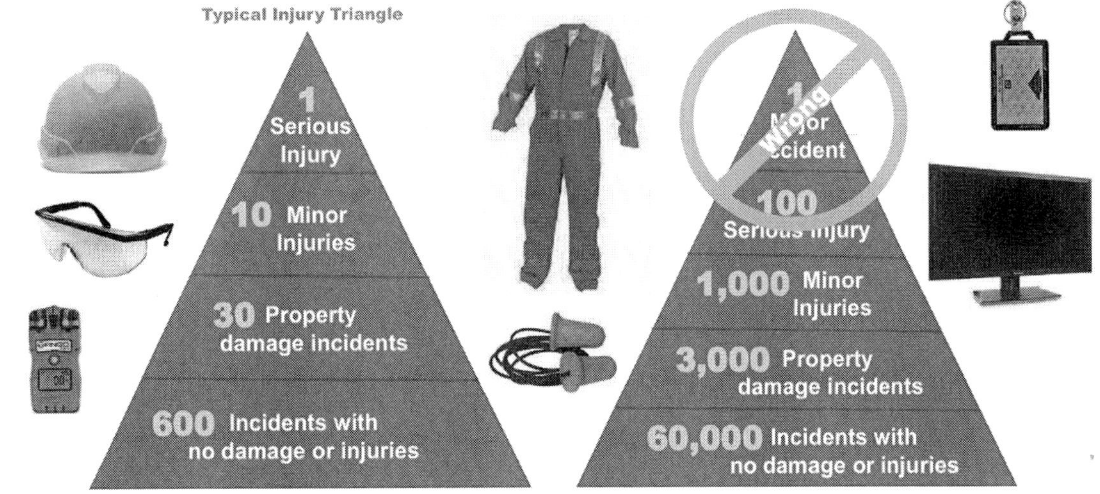

- Failure to Learn: the BP Texas City Refinery disaster. Andrew Hopkins 2010
- Conditions of occurrence of major and minor accidents, Institution of Occupational Safety and Health Journal 2001, 5(1): 7-21, Hale, A.

Two Triangle Model

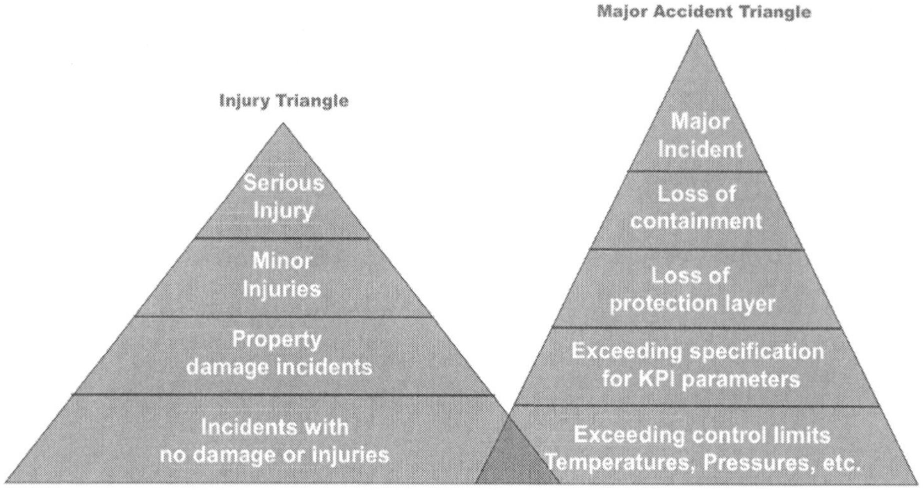

- Failure to Learn: the BP Texas City Refinery disaster, Andrew Hopkins 2010

INDUSTRY 4.0

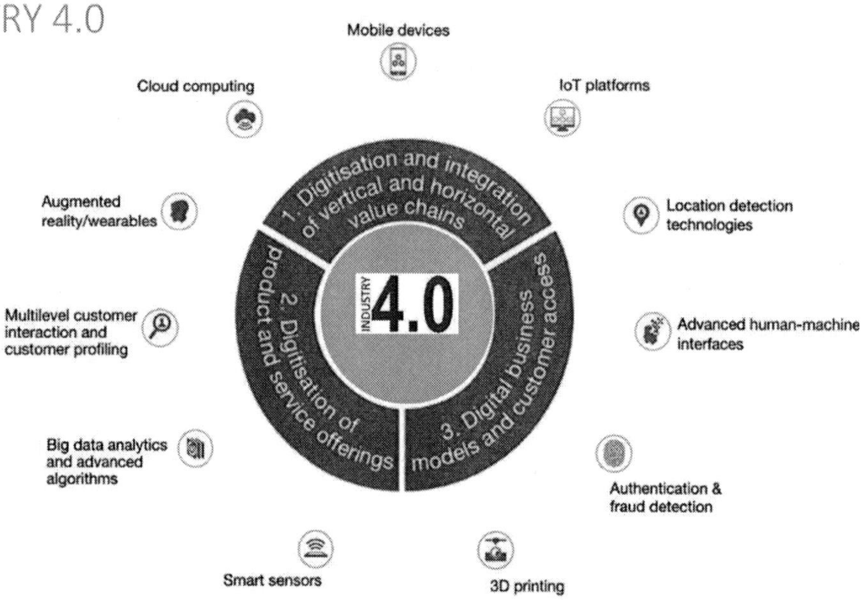

THE CORE OF INDUSTRY 4.0

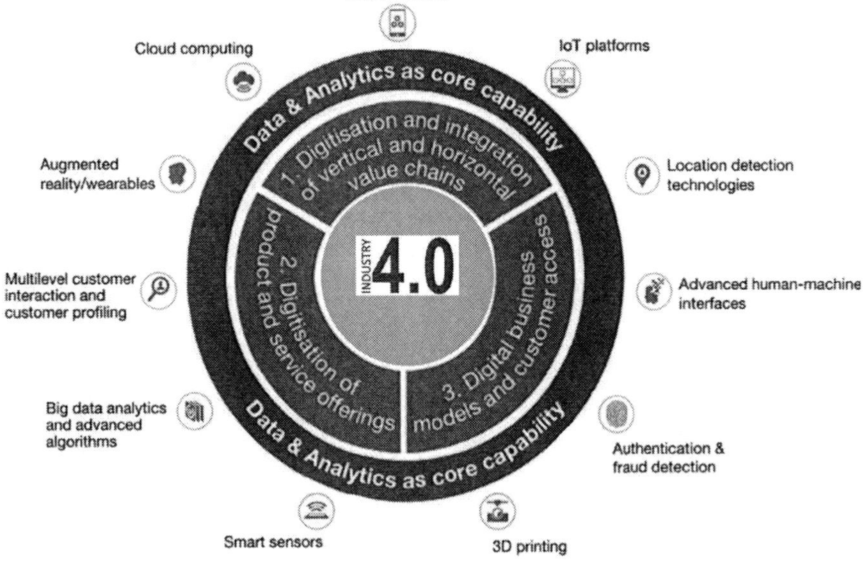

INDUSTRY 4.0

Cybernetic = *(Greek)* Skilled in steering

Connected Plant

Connected Worker

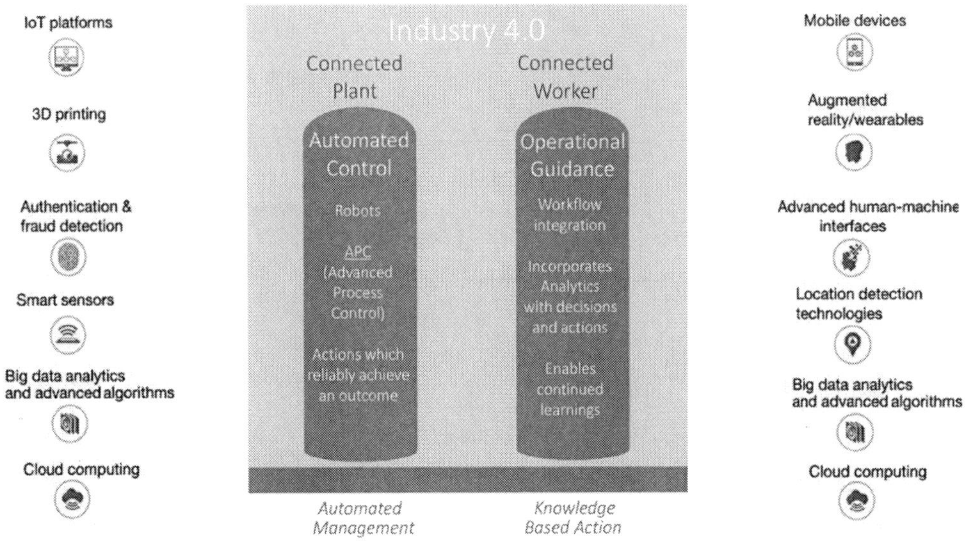

The Weakest Leg of the Stool

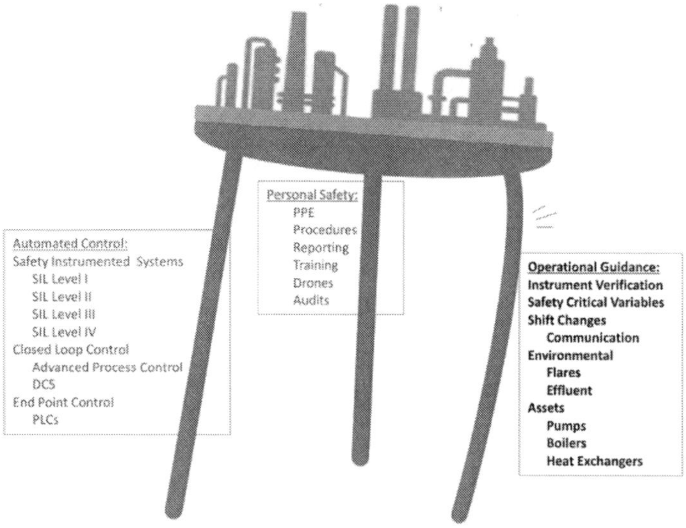

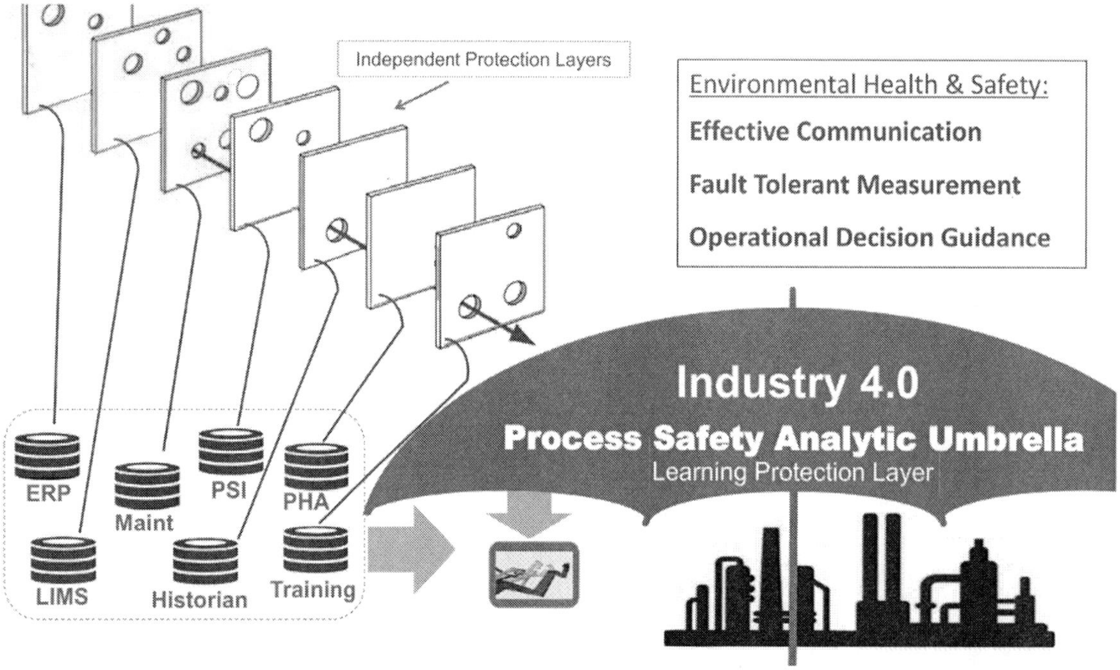

Texas City Refinery: Communication

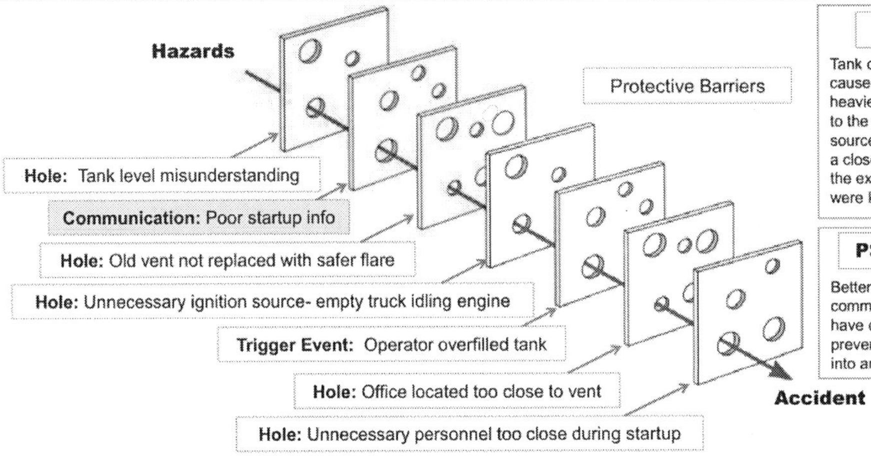

Consequences

Tank overfilled far beyond safe level, caused pressure relief vent to expel heavier than air chemicals which fell to the ground and found an ignition source. A meeting was being held in a close by temporary building when the explosion occurred. 15 people were killed, nearly 200 were injured.

PS 4.0 Opportunities

Better data aggregation, integrity, communication and guidance could have created protective barriers to prevent the situation from escalating into an incident.

Communication: Workers identified and reported various pieces of equipment on the column malfunctioning which were not rectified. *(CSB Report p.48).* Tank level instrument was reported as malfunctioning and had not been fixed. *(CSB Report pp. 319-323)*

Longford Gas Plant: Communication

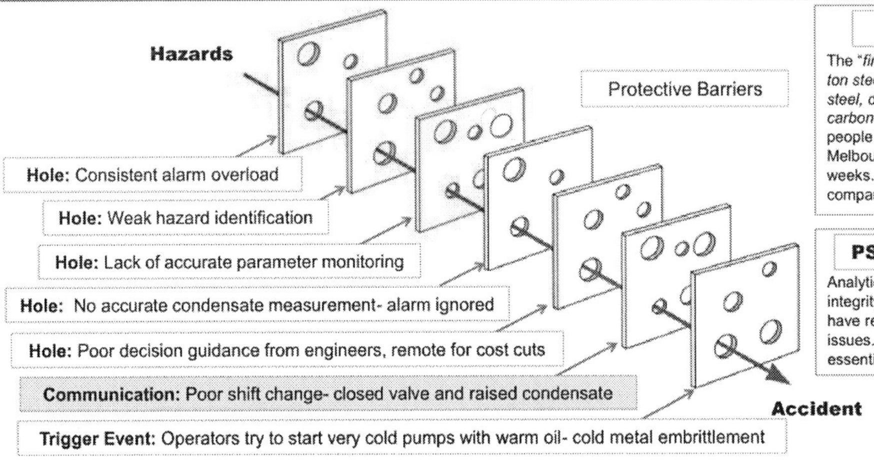

Consequences

The "*first explosion ripped apart a 14 ton steel vessel…it sent shards of steel, dust, debris and liquid hydro-carbon into the atmosphere.*" 2 people were killed, and 8 injured. Melbourne's gas supply was cut for 2 weeks. Litigation against the company reached over $1 billion.

PS 4.0 Opportunities

Analytic alarm filtering, measurement integrity, and communication would have reduced the risk for many safety issues. Decision guidance would be essential here.

Communication: Unaware of high condensate on previous shift, operator closed valve and raised condensate higher. (Dawson, 3.25-6)

Deep Water Horizon: Communication

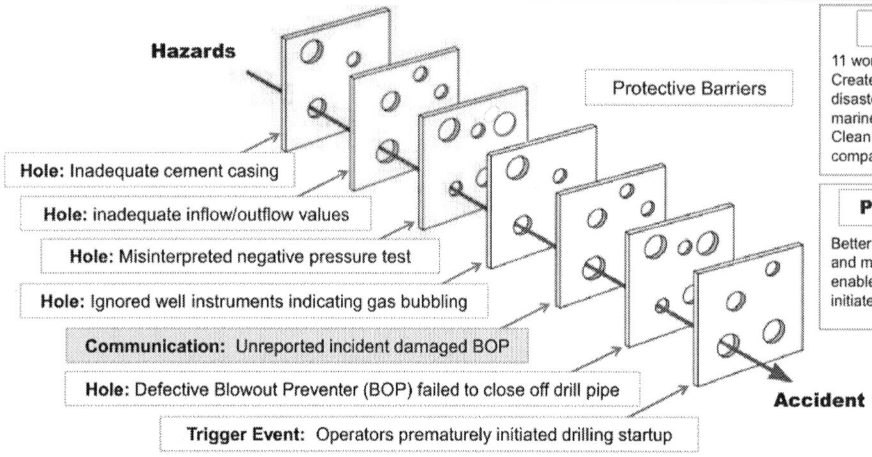

Consequences

11 workers were killed, and 17 injured. Created the largest environmental disaster in U.S. history and the largest marine accidental spill in the world. Clean up and litigation has cost the company over $54 billion USD.

PS 4.0 Opportunities

Better communication, data monitoring and measurement integrity would have enabled better decision guidance to not initiate a startup.

Communication: Knowledge of the damaged Blow out Preventer would likely have delayed the decision to drill.

Texas City Refinery: Measurement

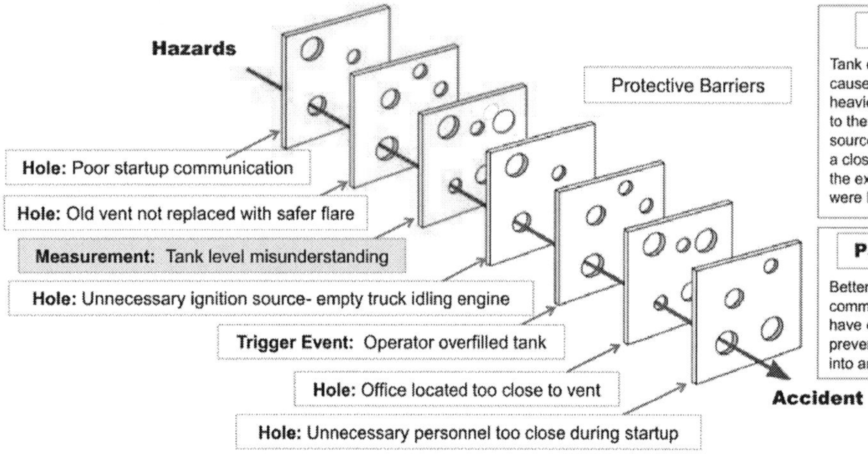

Consequences

Tank overfilled far beyond safe level, caused pressure relief vent to expel heavier than air chemicals which fell to the ground and found an ignition source. A meeting was being held in a close by temporary building when the explosion occurred. 15 people were killed, nearly 200 were injured.

PS 4.0 Opportunities

Better data aggregation, integrity, communication and guidance could have created protective barriers to prevent the situation from escalating into an incident.

Measurement: Instrument inaccuracy -reported the 52m splitter tower level at 2.4m when it was actually at 48m. Instrument could not be verified from physical level measurement as instrument was clouded and unreadable.

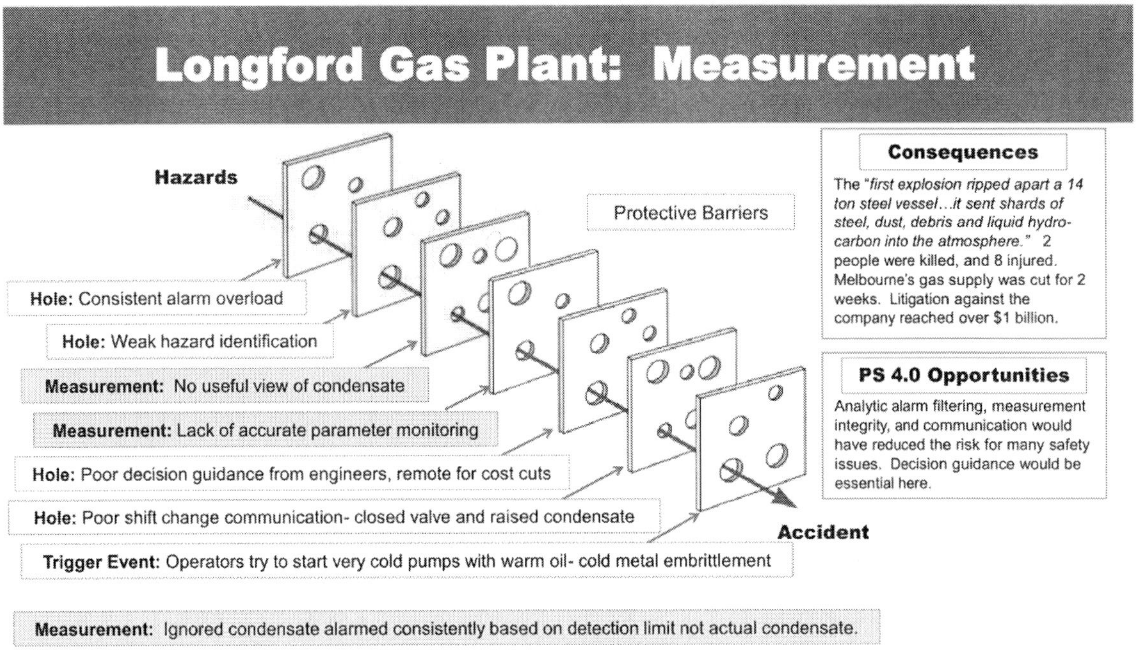
Longford Gas Plant: Measurement
Hazards
Protective Barriers
Consequences
The "first explosion ripped apart a 14 ton steel vessel…it sent shards of steel, dust, debris and liquid hydro-carbon into the atmosphere." 2 people were killed, and 8 injured. Melbourne's gas supply was cut for 2 weeks. Litigation against the company reached over $1 billion.
PS 4.0 Opportunities
Analytic alarm filtering, measurement integrity, and communication would have reduced the risk for many safety issues. Decision guidance would be essential here.
Hole: Consistent alarm overload
Hole: Weak hazard identification
Measurement: No useful view of condensate
Measurement: Lack of accurate parameter monitoring
Hole: Poor decision guidance from engineers, remote for cost cuts
Hole: Poor shift change communication- closed valve and raised condensate
Accident
Trigger Event: Operators try to start very cold pumps with warm oil- cold metal embrittlement
Measurement: Ignored condensate alarmed consistently based on detection limit not actual condensate.

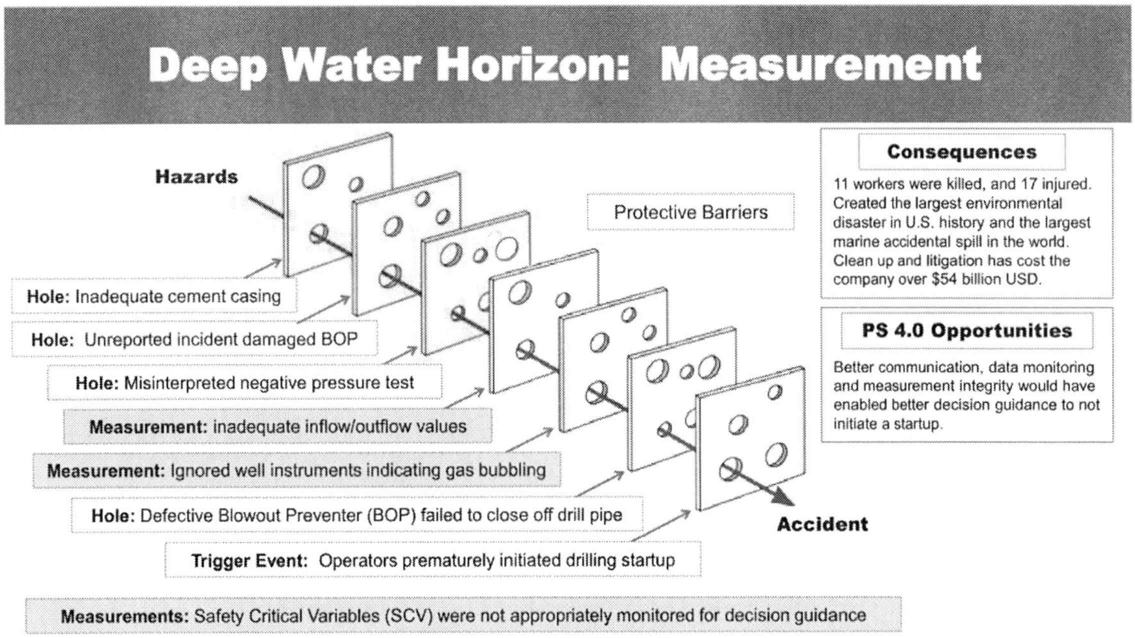
Deep Water Horizon: Measurement
Hazards
Protective Barriers
Consequences
11 workers were killed, and 17 injured. Created the largest environmental disaster in U.S. history and the largest marine accidental spill in the world. Clean up and litigation has cost the company over $54 billion USD.
PS 4.0 Opportunities
Better communication, data monitoring and measurement integrity would have enabled better decision guidance to not initiate a startup.
Hole: Inadequate cement casing
Hole: Unreported incident damaged BOP
Hole: Misinterpreted negative pressure test
Measurement: inadequate inflow/outflow values
Measurement: Ignored well instruments indicating gas bubbling
Hole: Defective Blowout Preventer (BOP) failed to close off drill pipe
Accident
Trigger Event: Operators prematurely initiated drilling startup
Measurements: Safety Critical Variables (SCV) were not appropriately monitored for decision guidance

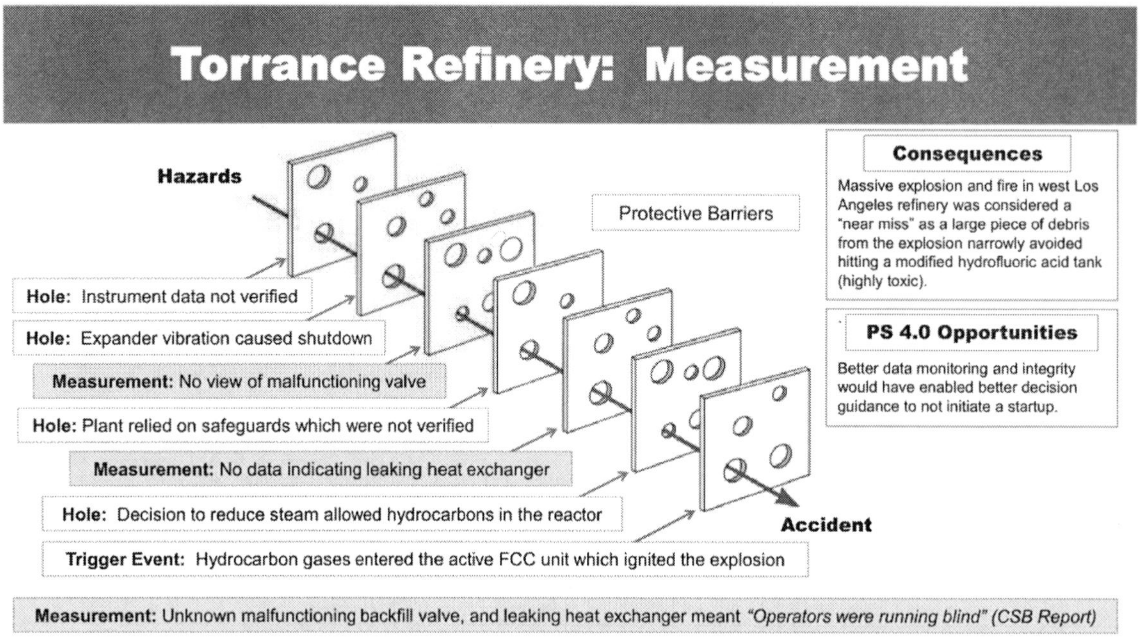
Torrance Refinery: Measurement
Hazards
Protective Barriers
Consequences
Massive explosion and fire in west Los Angeles refinery was considered a "near miss" as a large piece of debris from the explosion narrowly avoided hitting a modified hydrofluoric acid tank (highly toxic).
PS 4.0 Opportunities
Better data monitoring and integrity would have enabled better decision guidance to not initiate a startup.
Hole: Instrument data not verified
Hole: Expander vibration caused shutdown
Measurement: No view of malfunctioning valve
Hole: Plant relied on safeguards which were not verified
Measurement: No data indicating leaking heat exchanger
Hole: Decision to reduce steam allowed hydrocarbons in the reactor
Accident
Trigger Event: Hydrocarbon gases entered the active FCC unit which ignited the explosion
Measurement: Unknown malfunctioning backfill valve, and leaking heat exchanger meant "Operators were running blind" (CSB Report)

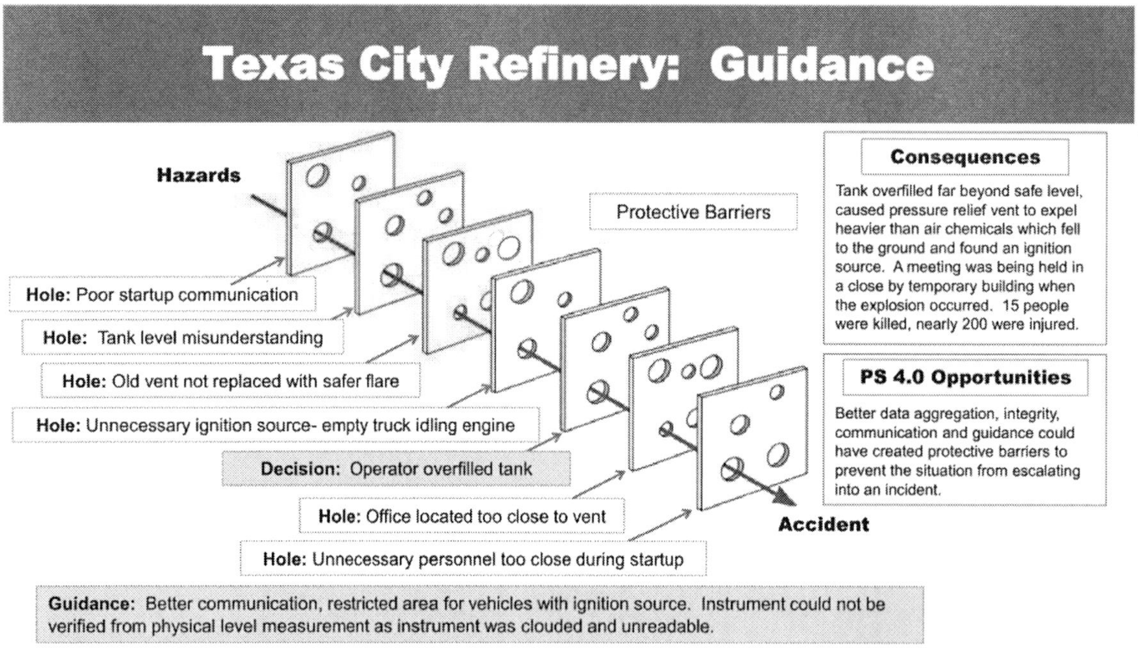
Texas City Refinery: Guidance
Hazards
Protective Barriers
Consequences
Tank overfilled far beyond safe level, caused pressure relief vent to expel heavier than air chemicals which fell to the ground and found an ignition source. A meeting was being held in a close by temporary building when the explosion occurred. 15 people were killed, nearly 200 were injured.
PS 4.0 Opportunities
Better data aggregation, integrity, communication and guidance could have created protective barriers to prevent the situation from escalating into an incident.
Hole: Poor startup communication
Hole: Tank level misunderstanding
Hole: Old vent not replaced with safer flare
Hole: Unnecessary ignition source- empty truck idling engine
Decision: Operator overfilled tank
Hole: Office located too close to vent
Accident
Hole: Unnecessary personnel too close during startup
Guidance: Better communication, restricted area for vehicles with ignition source. Instrument could not be verified from physical level measurement as instrument was clouded and unreadable.

Torrance Refinery: Guidance

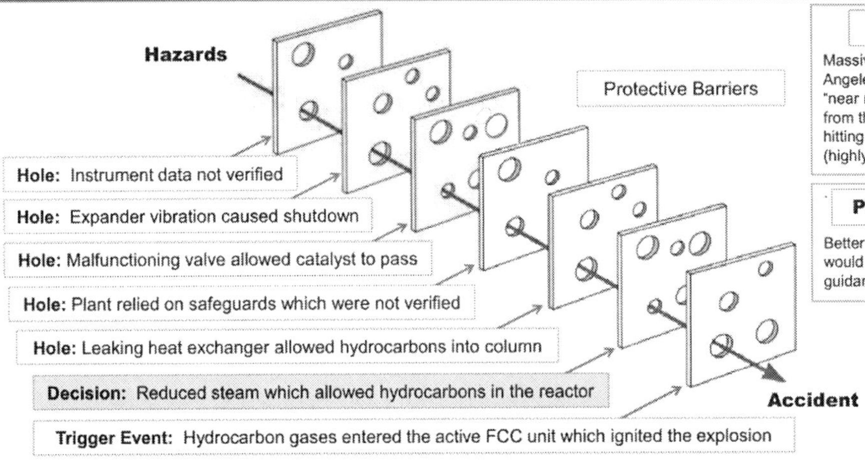

Consequences

Massive explosion and fire in west Los Angeles refinery was considered a "near miss" as a large piece of debris from the explosion narrowly avoided hitting a modified hydrofluoric acid tank (highly toxic).

PS 4.0 Opportunities

Better data monitoring and integrity would have enabled better decision guidance to not initiate a startup.

Guidance: Operations (blind to data) reduced steam to repair expander, which allowed hydrocarbons to enter the FCC unit. Closed catalyst backfill valve was faulty and hydrocarbons were still coming into the chamber

Deep Water Horizon: Guidance

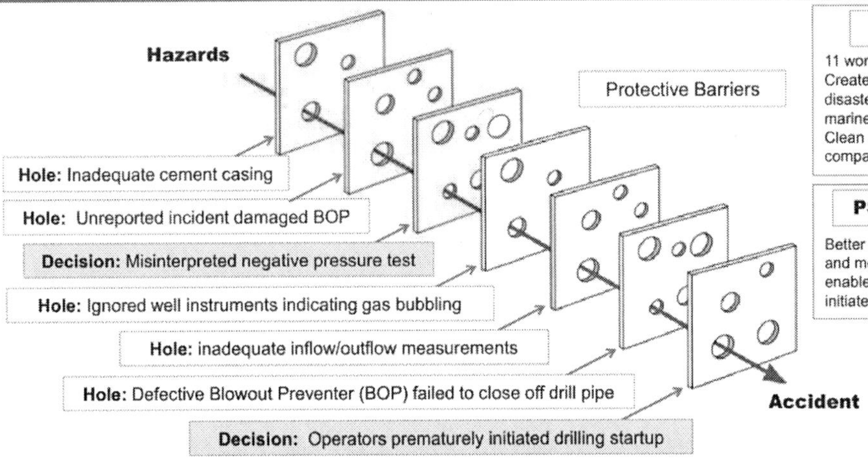

Consequences

11 workers were killed, and 17 injured. Created the largest environmental disaster in U.S. history and the largest marine accidental spill in the world. Clean up and litigation has cost the company over $54 billion USD.

PS 4.0 Opportunities

Better communication, data monitoring and measurement integrity would have enabled better decision guidance to not initiate a startup.

Guidance: Poor analysis of negative pressure test and gas bubbling instruments led to premature decision to initiate drilling

Example KPI with Analytic Monitoring

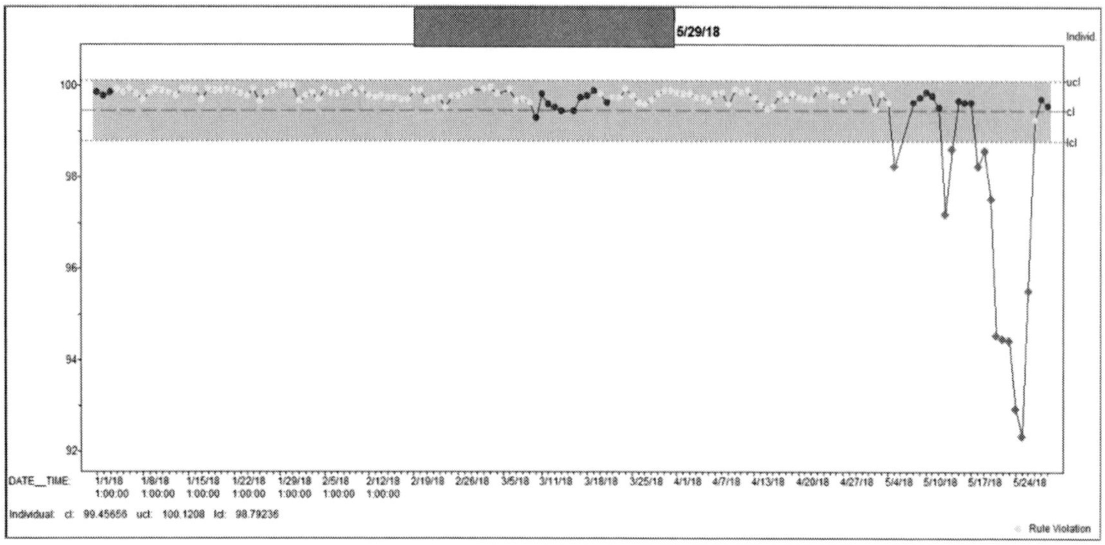

Example Parameter

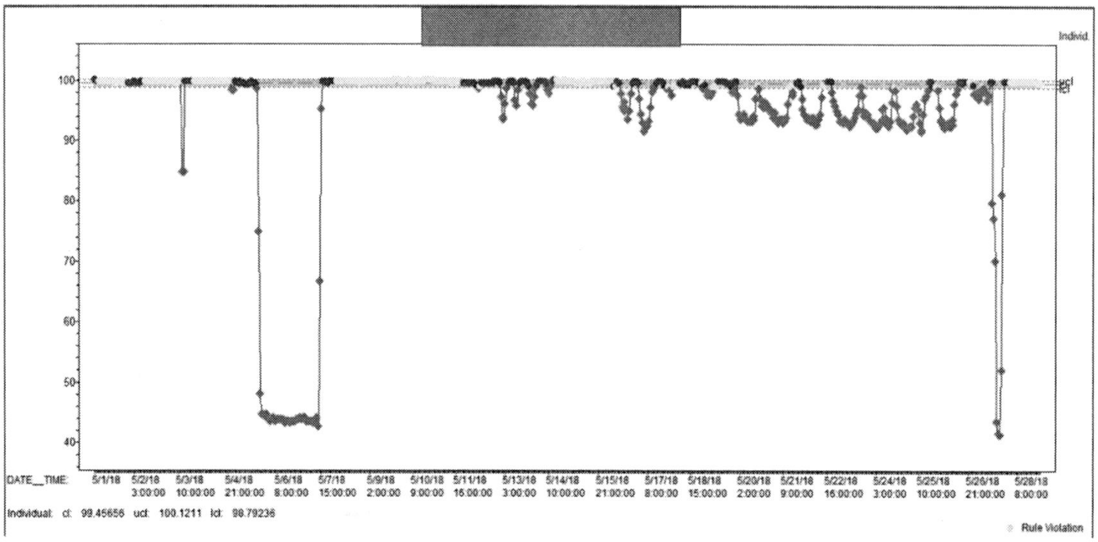

		As of March 31, 2017		As of March 31, 2018
Assets				
Current assets				
Cash and deposits	(*2)	556,836	(*2)	571,911
Notes and accounts receivable - trade		164,761		176,020
Merchandise and finished goods		118,195		143,332
Work in process		19,695		21,901
Raw materials and supplies		26,139		38,551
Short-term loans receivable		1,766		806
Deferred tax assets		8,626		19,426
Other	(*2)	76,590	(*2)	95,542
Allowance for doubtful accounts		(1,209)		(1,229)
Total current assets		971,401		1,066,264
Non-current assets				
Property, plant and equipment				
Buildings and structures, net		77,002		89,314
Machinery, equipment and vehicles, net		100,516		113,266
Tools, furniture and fixtures, net		22,349		46,237
Land		89,925		94,565
Construction in progress		19,210		33,698
Total property, plant and equipment	(*1), (*2)	309,004	(*1), (*2)	377,082
Intangible assets		22,939		29,022
Investments and other assets				
Investment securities	(*2), (*3)	94,959	(*2), (*3)	97,699
Long-term loans receivable		21,615		6,218
Net defined benefit asset		2,913		4,198
Deferred tax assets		7,311		10,558
Other	(*2), (*3)	60,120	(*2), (*3)	70,178
Allowance for doubtful accounts		(5,853)		(5,921)
Total investments and other assets		181,068		182,931
Total non-current assets		513,011		589,035
Total assets		1,484,413		1,655,299

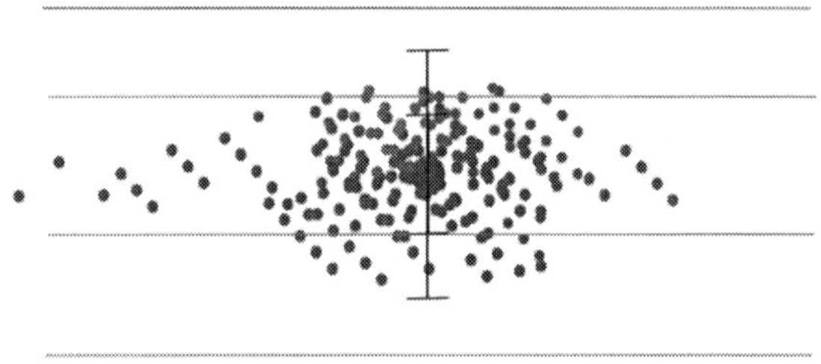

Data Interrelationships:

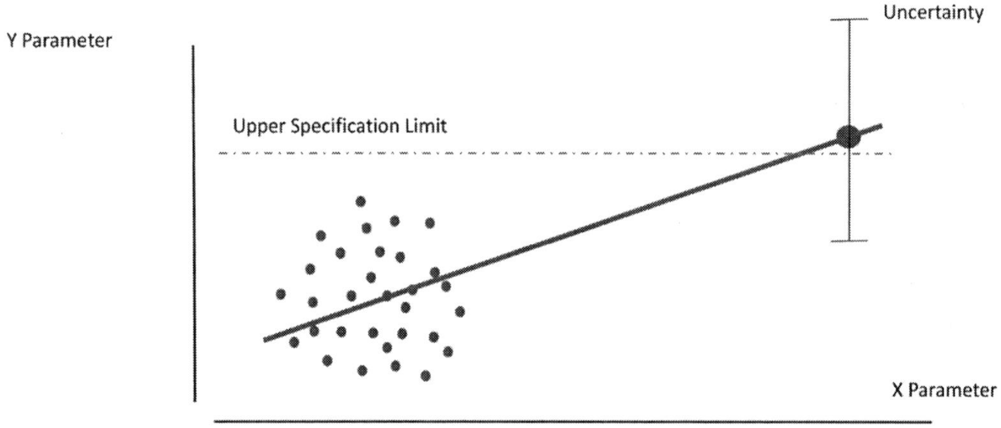

First example:

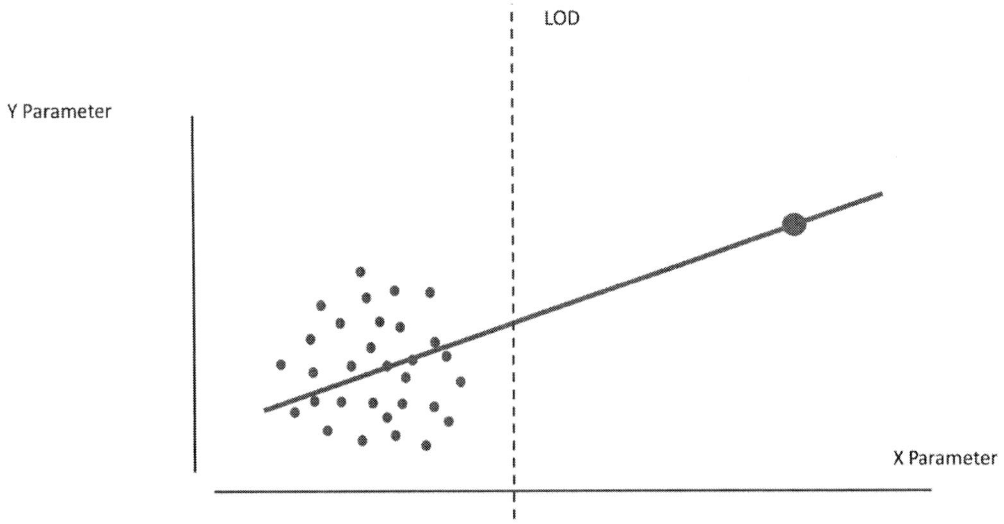

Constructing a Story with Data

Putting the data in context:

- **The Setting:** Where does the story take place, and what is the timeline?
- **Hero / Main character:** Who is driving the action?
- **When does imbalance occur:** What is driving our main character to action?
- **The Balance:** What do you want to see happen?
- **The Solution:** What is the outcome?

EHS Actionboards

Dashboards:

- Unclear what to do
- Effectiveness varies
- Inconsistent reviews

Actionboards:

- Clear course of action
- Consistent effective actions
- Real-time responsiveness

Environmental Health & Safety:

- **Effective Communication**
- **Fault Tolerant Measurement**
- **Operational Decision Guidance**

Thank You!

Contact:

Jim Petrusich
+1-503-224-7727
jpetrusich@nwasoft.com
www.nwasoft.com

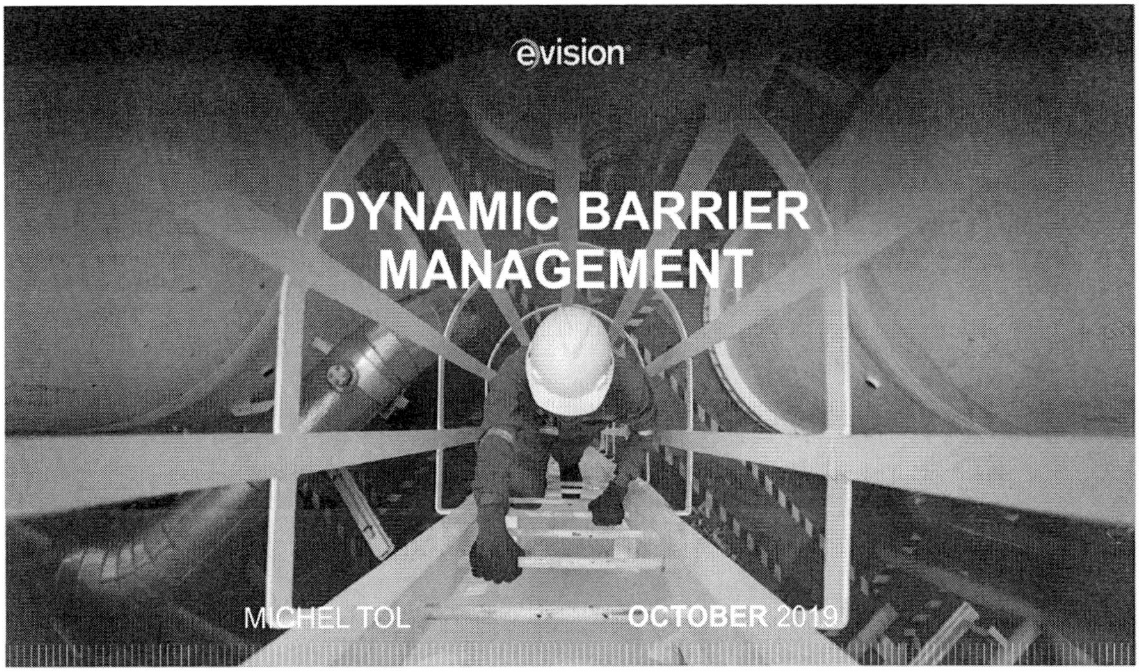

- World leader in digital Control of Work
- Rapid innovator, aspiring full COW, EHS and PSM domain coverage
- Multi-vertical approach: O&G, (petro)chemical, pharma, and more
- Highly skilled industry experts, consultants, developers and leaders
- Dedicated to helping you achieve Excellence
- A Wolters Kluwer Business

evision

Project Evaluation

- A records 1000+ trained users in 9 months.
- Embedding the business rules in the system led to 50% reduction in permit procedures and work instructions.
- Permit template feature helped in standardising repeated jobs and also saved time for drafting.
- It enforced more discipline and accountability
- Based on the positive outcome, the further rollout for the Electronic Permit to Work system is also planned in other locations of Borouge

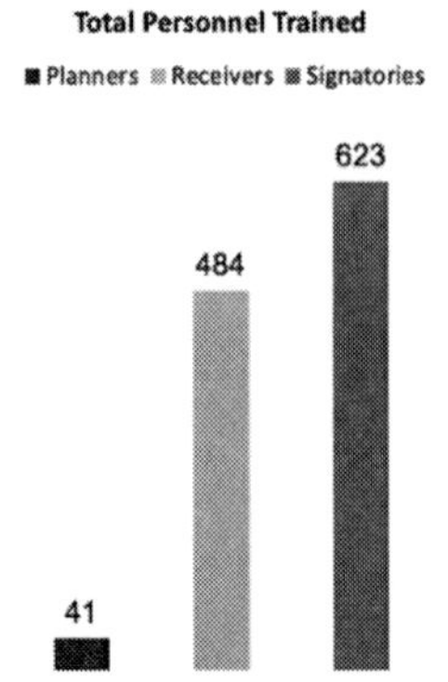

Borouge

eVision

Safety Moment – Shell Sherwood Terminal

Incident ID:	2243856 Review for LFI Add Action Item
Status:	Closed
Responsible Department/Unit:	Sherwood Terminal : STO/37 - Dist Ops Canada:Sherwood:Terminal
Event Type:	Unsafe Act/Condition (Potential Incident)
Incident Sub Types:	Unsafe Act/Condition (Potential Incident): Unsafe Act - People
Responsible Supervisor:	Hamel, Olivier
Short Description:	PI - Contractor not prepared to work near Microwave tower
Responsibility:	Incident Owner: Hamel, Olivier Reported By: Hamel, Olivier

Hey guys, a good news story from the new EVision software, this probably would not have been caught under the old permitting system. At first it was seen as a impediment to getting the permit issued for work we thought was ready to go, but it become apparent the contractor was not aware of a certain requirements prompted by the software. Kudos to Jim and the operator for pausing the process until resolution could be found, and thanks Erin for the quick support that morning.

Olivier Hamel
Sherwood Terminal Manager
Shell Canada Products

Contractor (Tridon) was not well enough prepared to do work near the microwave tower which has been identified as a source of radiation. During the permit writing process, the EVision software prompted a question about working near radiation that the contractor was not able to successfully answer or show proper documentation per the control framework. Work was stopped, discussion ongoing with HSE to determine the proper requirements before work can be rescheduled.

eVision

INDUSTRY ?.0 ...?

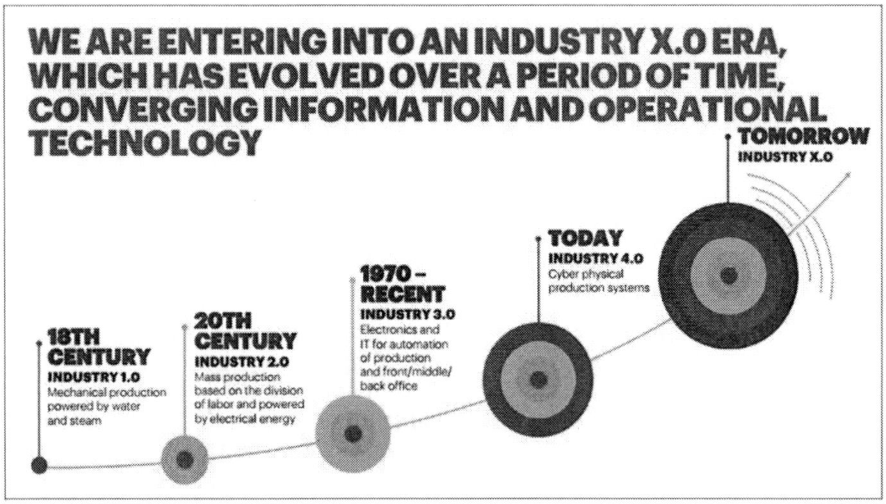

WHAT OUR CLIENTS **DESIRE**

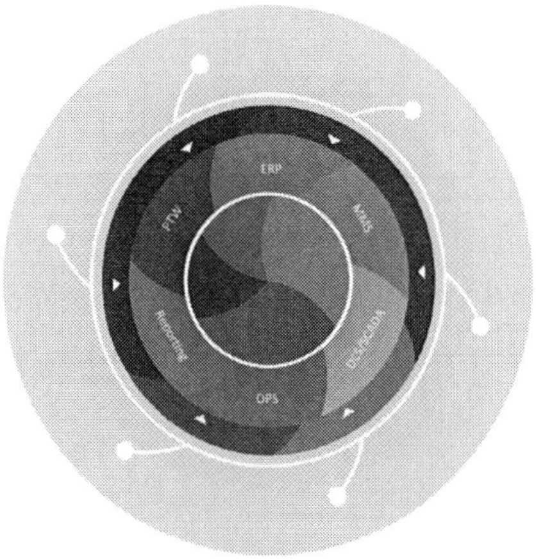

THE IMPORTANCE OF **PLATFORM THINKING**

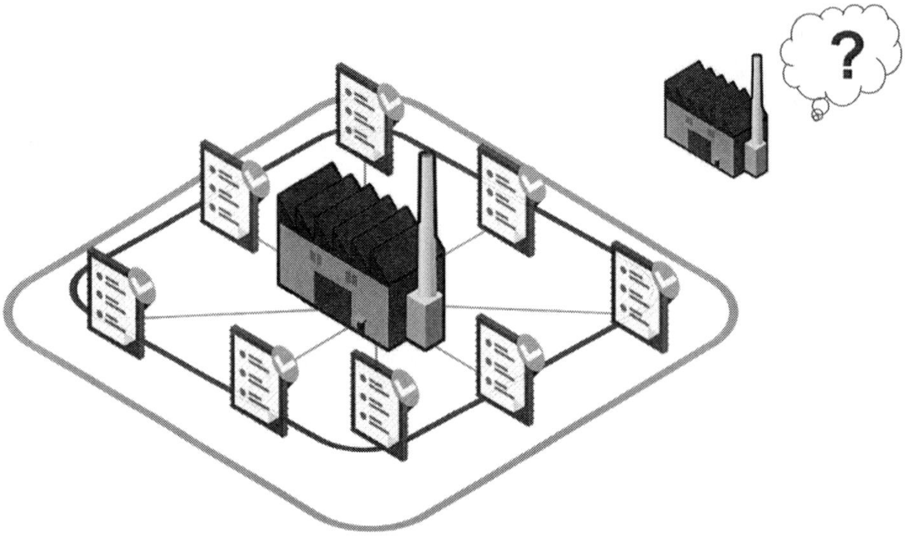

A barrier is a risk control that seeks to **prevent unintended events** from occurring or **prevent escalation of events** into accidents with harmful consequences

WHY BARRIER MANAGEMENT IS **IMPORTANT**

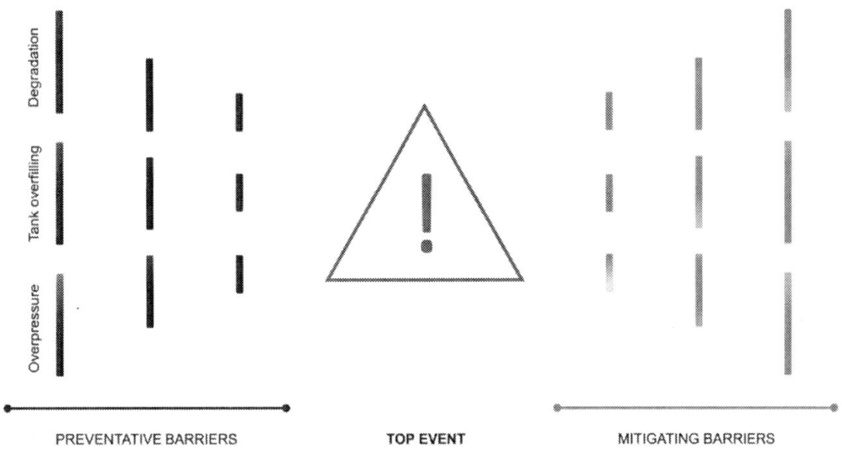

WHY BARRIER MANAGEMENT IS **IMPORTANT**

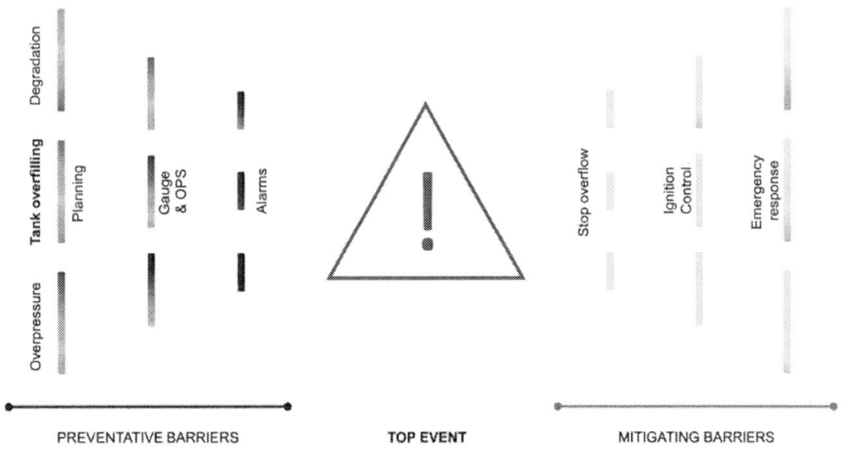

CORE OF THE PROBLEM

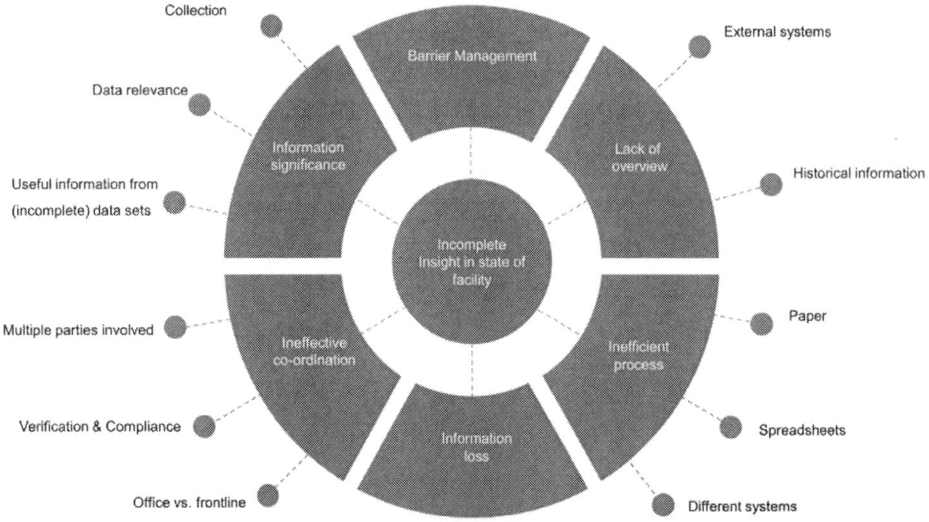

11

BARRIER VISION – **SWISS-CHEESE OVERVIEW**

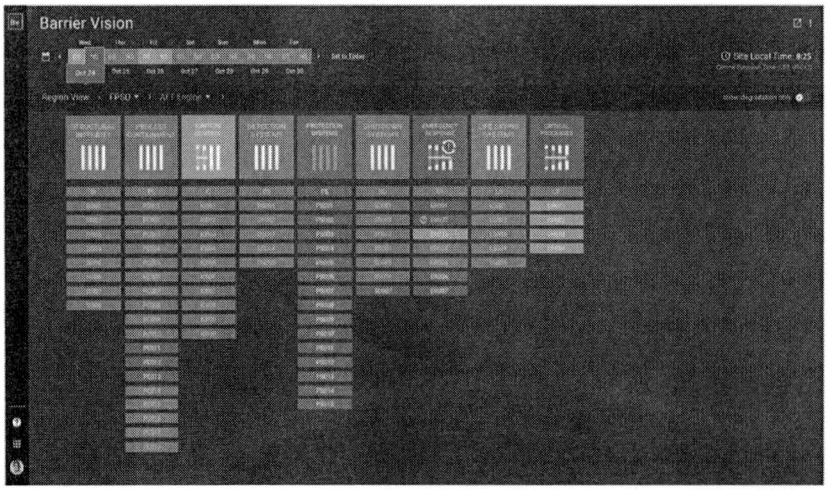

12

Barrier Vision – **Swiss-Cheese overview**

Barrier detail overview

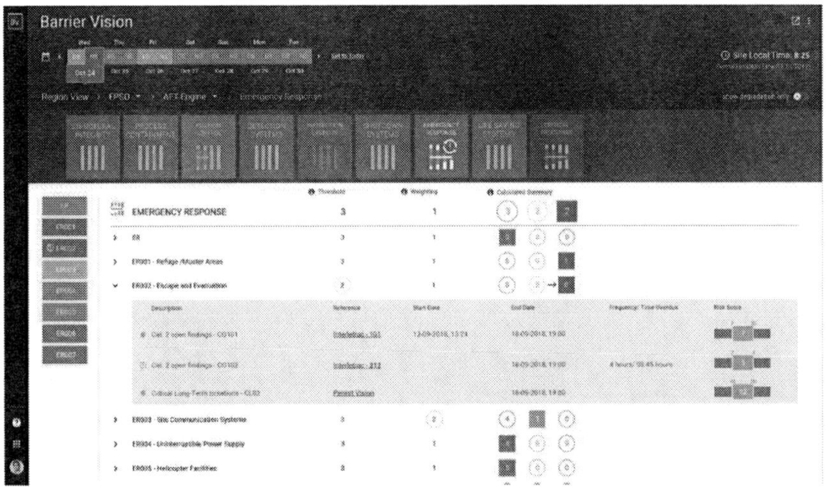

REAL-LIFE SCENARIO: PERMIT TO WORK MEETS BARRIER MANAGEMENT

THANK YOU
Learn more at
www.evision-software.com

Research and Education in Process and Plant Safety

Center for Chemical Process Safety (CCPS) 2019

Overcoming of accuracy and computing time of online model-based reactor safety concepts

Johannes Biernath, M.Sc

CSE-Institute – Research and Innovation in Process- and Plant Safety   johannes.biernath@cse-institut.de

Johannes Biernath, M.Sc.

Education	Chemical / Process Engineering	
2014	**Partical semester**	Thesis: Simulation of thin film flows in columns
2015	**Master of Science**	Thesis: Development of Experimental Setup for Investigation of Depositions in Rod Bundels
Research 2016	Research Assistant	Tank breathing, Pipeline Security
PhD 2017	SmartHIP	Smart Technology for High Integrity Process Control

How to protect against runaway reaction?

- **Online model-based** reactor safety concept
- **Real-time** process data acquisition
- Implementation of models in failsafe programmable logic controller (FPLC)
- **Self-regulation** for reactor safety

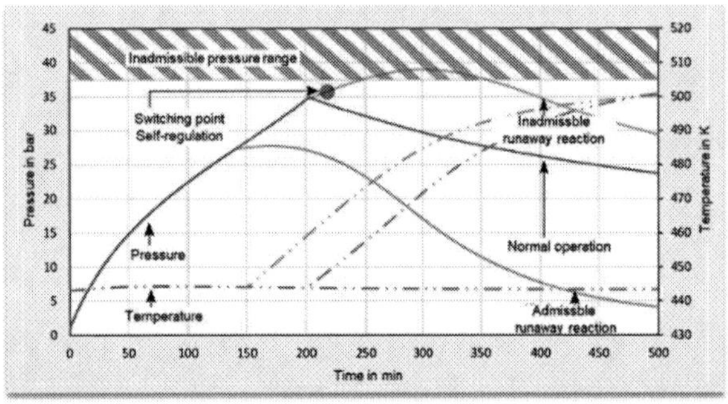

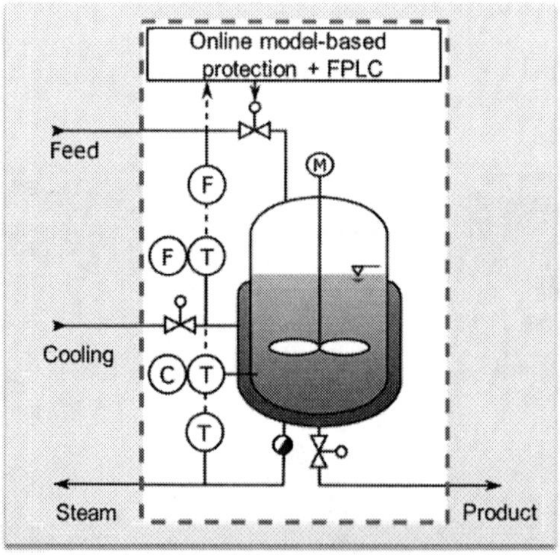

Model-based reactor safety concepts

- **Reduction of accuracy by**
  - Process (e.g. inhomogeneity, viscosity, impurities)
  - Software (e.g. arithmetic, real-time capabilities, programmcode)
  - Hardware (e.g. measured value acquisition and processing, reliability)

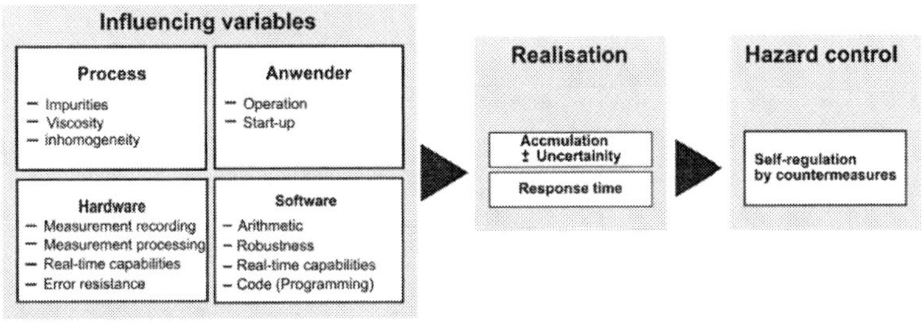

Data fusion – Combination of safety models

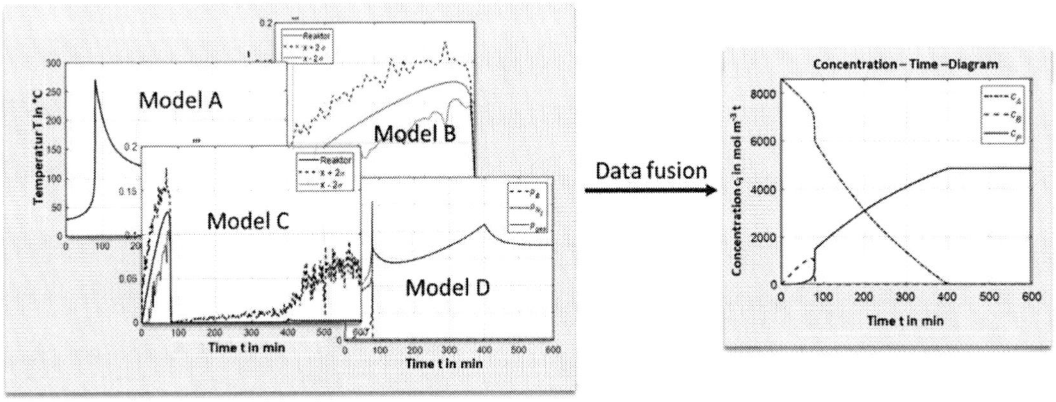

Dempster-Shafer theory

- Definiton of a **perceptual framework** $\boldsymbol{\Theta}$:

$$\Theta = \{Safety, Danger\}$$

- Subdivision into **hypothesis** $\mathbf{A} \subseteq \boldsymbol{\Theta}$:

$$A = \{safe, dangerous, safe \cup dangerous, empty\ set\}$$

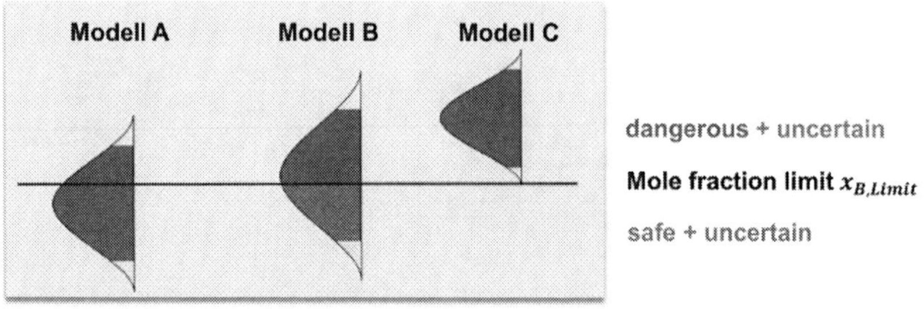

Dempster-Shafer theory

- Model A – normal condition

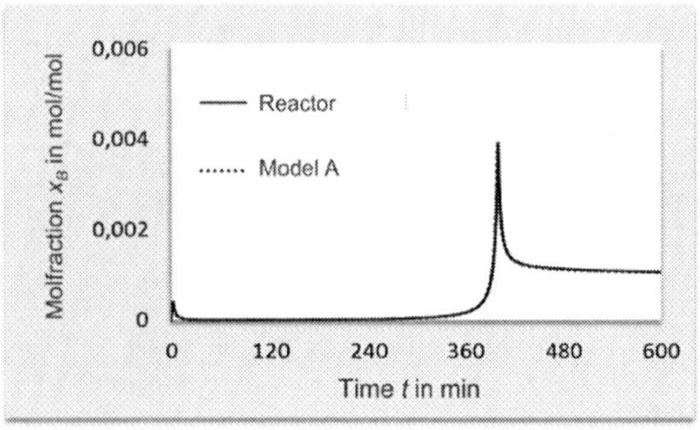

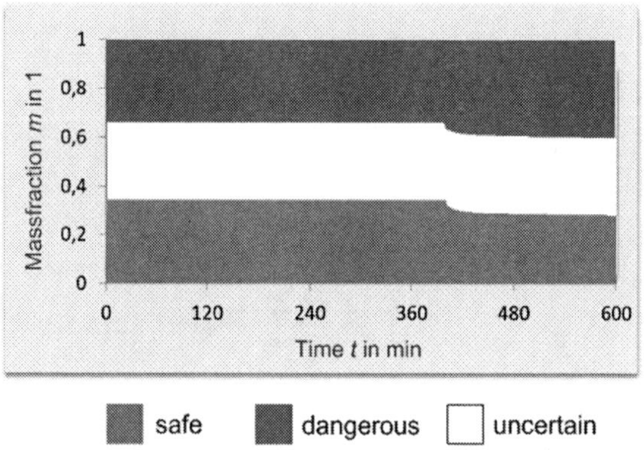

Dempster-Shafer theory

- Model B – normal condition

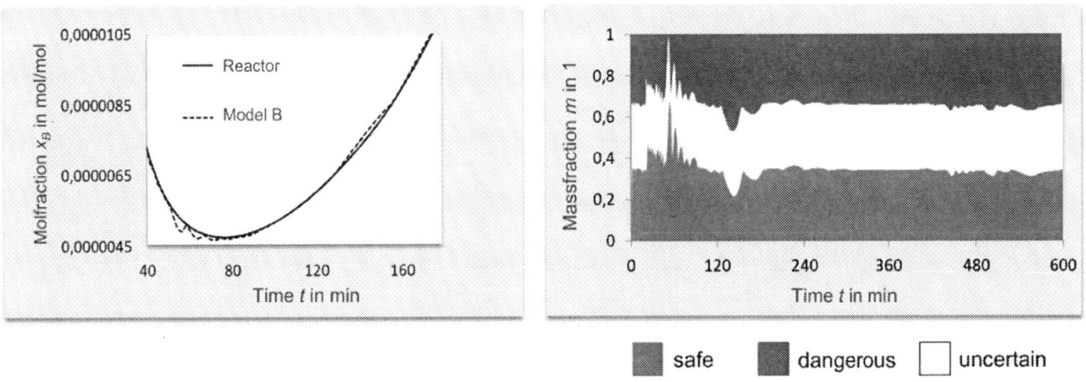

Dempster-Shafer theory

- Model C – normal condition

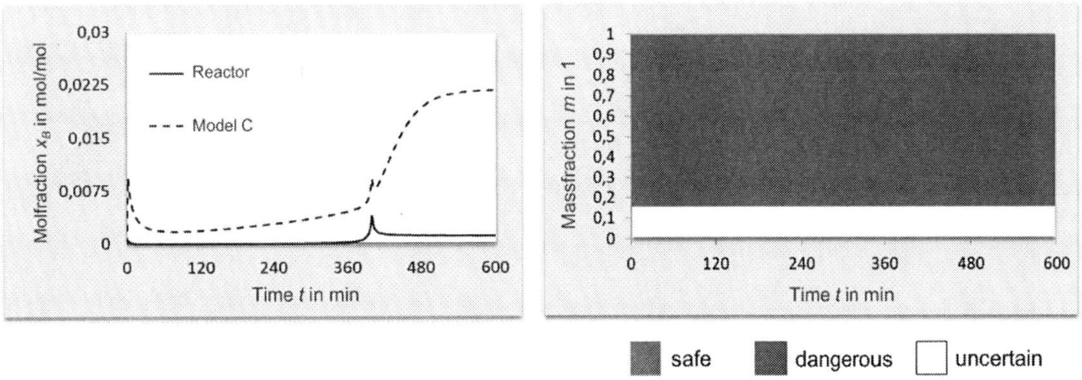

Dempster-Shafer theory

- **Results**:
  - Increasing the accuracy by reducing the uncertainty of each single safety model
  - Improving the availability due to the combination
  - Dempster Shafer algorithm valued not suitable safety models

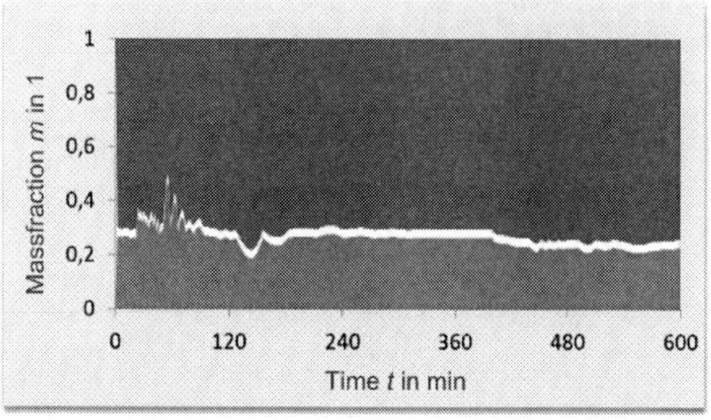

Summary

- Reduction of accuracy by **process**, **hardware** and **software**
- **Combination** of models increases accuracy and availability by **Dempster Shafer**

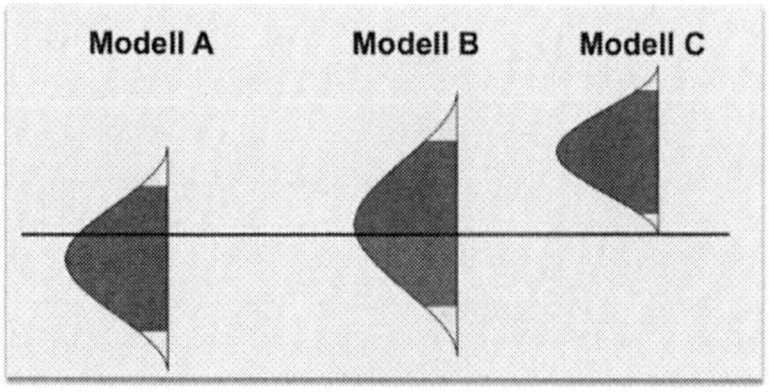

From Incidents to Intelligent KPIs: The construction of safety indexes based on events

Chabane MAZRI, Stephen QUEVA

02/10/2019

Chabane.mazri@ineris.fr

INERIS

maîtriser le risque
pour un développement durable

I. Context and objectives

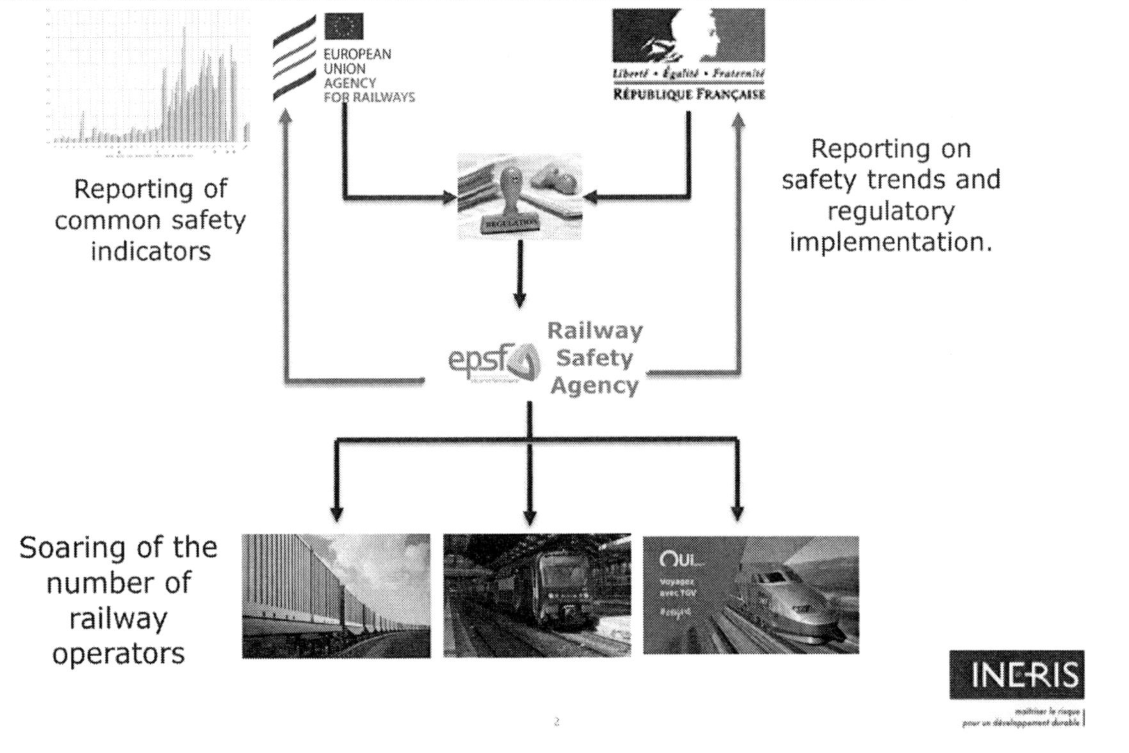

INERIS

maîtriser le risque
pour un développement durable

2

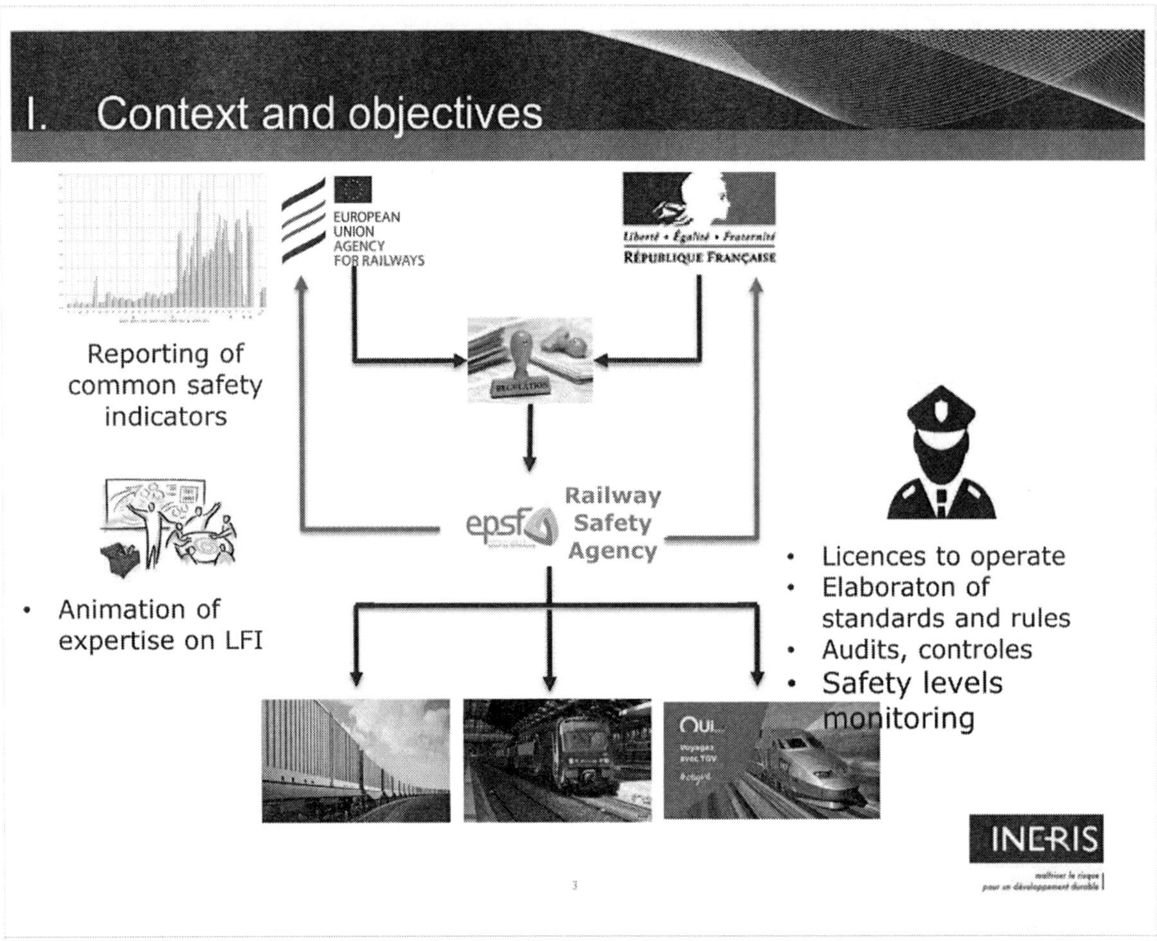
I. Context and objectives
EUROPEAN UNION AGENCY FOR RAILWAYS
Liberté • Égalité • Fraternité
RÉPUBLIQUE FRANÇAISE
Reporting of common safety indicators
• Animation of expertise on LFI
epsf
Railway Safety Agency
• Licences to operate
• Elaboraton of standards and rules
• Audits, controles
• Safety levels monitoring
OUI...
INERIS
3

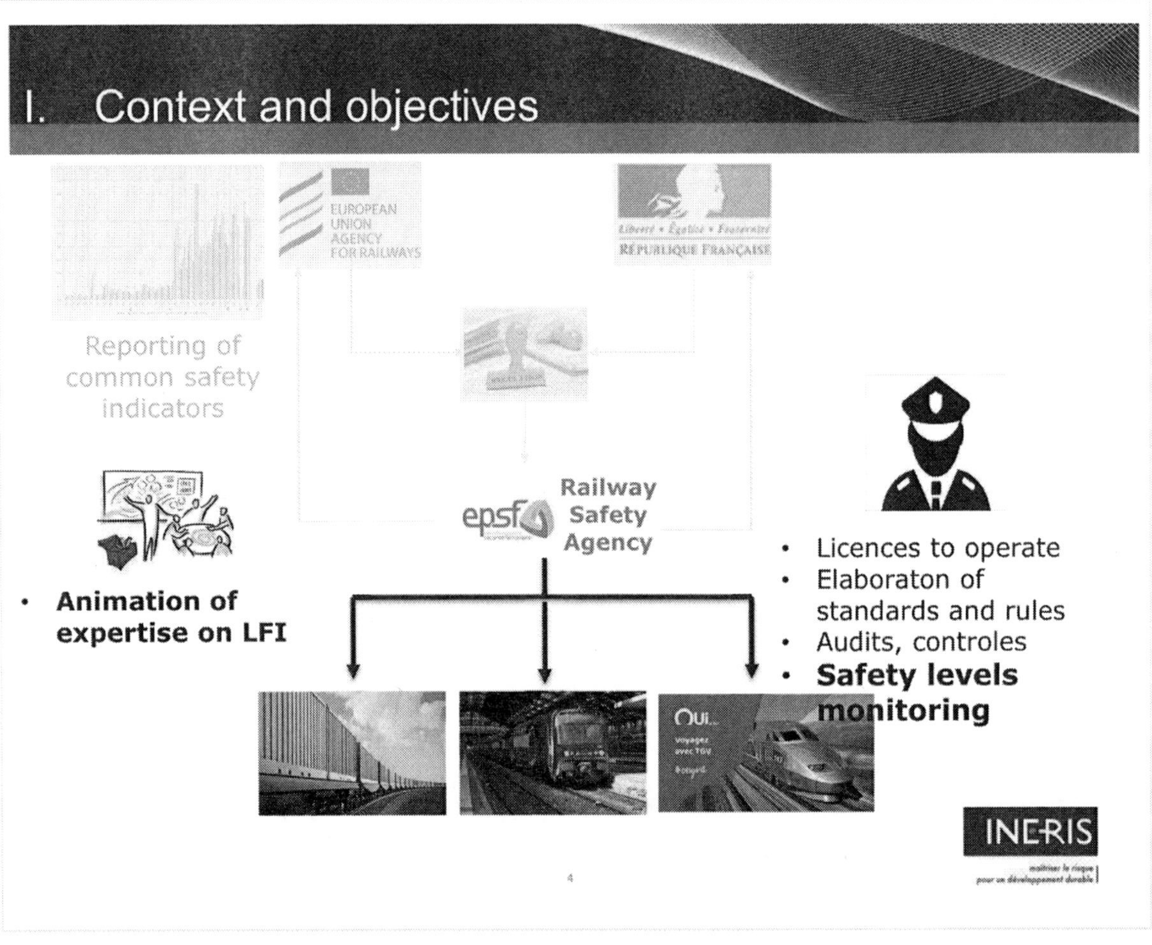
I. Context and objectives
EUROPEAN UNION AGENCY FOR RAILWAYS
Liberté • Égalité • Fraternité
RÉPUBLIQUE FRANÇAISE
Reporting of common safety indicators
• Animation of expertise on LFI
epsf
Railway Safety Agency
• Licences to operate
• Elaboraton of standards and rules
• Audits, controles
• Safety levels monitoring
OUI...
INERIS
4

I. Context and objectives

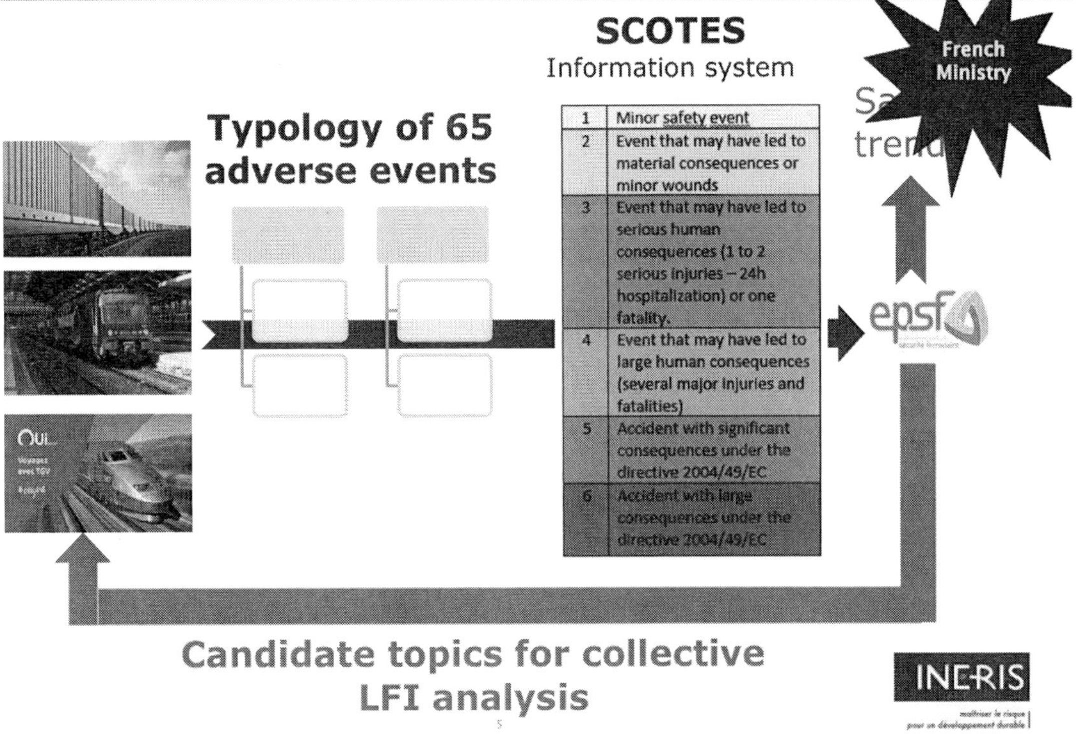

INERIS
maîtriser le risque | pour un développement durable

5

II. Methodological challenges

Given the incidents and accidents occurrences, which events deserve our attention for further investigation in collective LFI sessions?

Formulation 1

	L1	L2	L3	L4	L5	L6	Priority ranking of event $_i$
Safety event $_i$	N_1	N_2	N_3	N_4	N_5	N_6	**G (N_1, N_2, N_3, N_4, N_5, N_6)**

INERIS
maîtriser le risque | pour un développement durable

6

II. Methodological challenges

Given the incidents and accidents occurrences, what are safety trends and how are they evolving?

Formulation 2

	L1	L2	L3	L4	L5	L6	Level of risk posed by each event
Safety event $_i$	N_{1i}	N_{2i}	N_{3i}	N_{4i}	N_{5i}	N_{6i}	**$G(N_{1i}, N_{2i}, N_{3i}, N_{4i}, N_{5i}, N_{6i})$**
Safety event $_j$	N_{1j}	N_{2j}	N_{3j}	N_{4j}	N_{5j}	N_{6j}	**$G(N_{1j}, N_{2j}, N_{3j}, N_{4j}, N_{5j}, N_{6j})$**
....	.....	....	....	....	....	....	
Global safety level	**$\sum N_1$**	**$\sum N_2$**	**$\sum N_3$**	**$\sum N_4$**	**$\sum N_5$**	**$\sum N_6$**	**$G(\sum N_1, \sum N_2, \sum N_3, \sum N_4, \sum N_5, \sum N_6)$**

INERIS
maîtriser le risque | pour un développement durable |

7

III. Key models

- **Do not wait for accidents to acknowledge a decline in safety levels (proactive abilities).**
  - 0 accidents ≠ good safety levels.
  - More accidents / incidents ≠ A decrease in safety performances (weak statistical representativeness, aleatory element)
- **At EU level, France belongs to the leading group in terms of safety performances.**
  - Very few accidents.
  - Incidents become a key source of knowledge for safety improvement.

INERIS
maîtriser le risque | pour un développement durable |

8

III. Key models

- **No aggregation of incidents and accidents.**
  - Cognitive challenge of comparing incidents and accidents.
  - A unique index will hardly reflect modification in number or intensity of incident occurrences.
  - Separate aggregation of incidents and accidents is therefore required
    - IC = (L_1, L_2, L_3, L_4)
    - AC = (L_5, L_6)

1	Minor safety event
2	Event that may have led to material consequences or minor wounds
3	Event that may have led to serious human consequences (1 to 2 serious injuries – 24h hospitalization) or one fatality.
4	Event that may have led to large human consequences (several major injuries and fatalities)
5	Accident with significant consequences under the directive 2004/49/EC
6	Accident with large consequences under the directive 2004/49/EC

INERIS

9

IV. Safety indexes design

Defining LFI priorities

Safety frontier

LFI should focus primarily on events displaying the worst combination of IC and AC records.

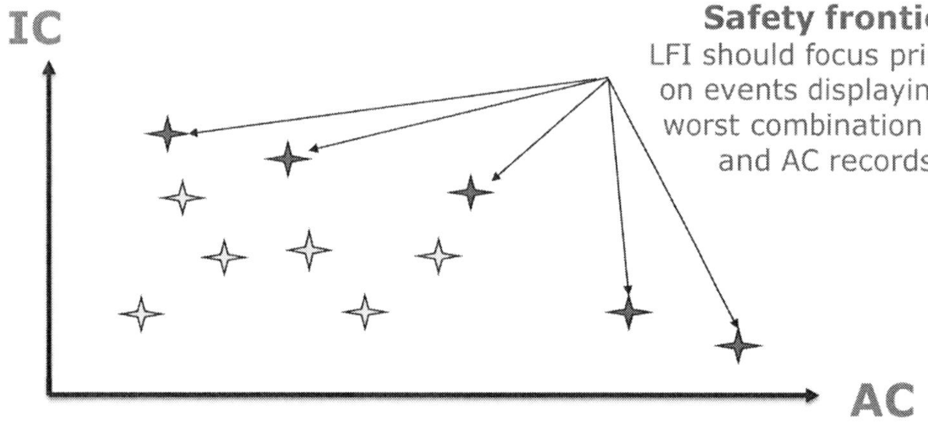

INERIS

10

IV. Safety indexes design

Safety trends

- Each time period should be represented by a composite index

$$GSL = (IC_G, AC_G)$$

Pays	Note	Perspective	Date	Source
Albanie	B+	Stable	juin 2011	3
Angola	B+	Stable	juin 2011	3
Argentine	SD	Négative	février 2016	4
Aruba	A-	Négative	juin 2011	5
Australie	AAA	Stable	juin 2011	5
Autriche	AA+	Négative	février 2012	5
Azerbaïdjan	BB+	Positive	juin 2011	5
Bahamas	BBB+	Stable	juin 2011	5
Bahreïn	BBB	Négative	juin 2011	5
Bangladesh	BB-	Stable	juin 2011	5

INERIS
maîtriser le risque | pour un développement durable |

11

IV. Safety indexes design

- To synthesize, both events classification and safety trends require for AC and IC to be designed.
  - $IC = W_1 N_1 + W_2 N_2 + W_3 N_3 + W_4 N_4$
  - $AC = W_5 N_5 + W_6 N_6$
- New problem formulation:
  - Weights design of (W_1, W_2, W_3, W_4) and (W_5, W_6)

INERIS
maîtriser le risque | pour un développement durable |

12

IV. Safety indexes design

Weights design

- Cohérence and legitimacy issues.
- **Methodological framework**
  1. AHP (Analytic Hierarchical process) developped by Saaty (1972).
  2. Vast array of applications: Military, logistics, environment…and risk analysis
- **Key features**: Allow DMs toexpress preferences in a "natural way" before transforming them into normalized weights.

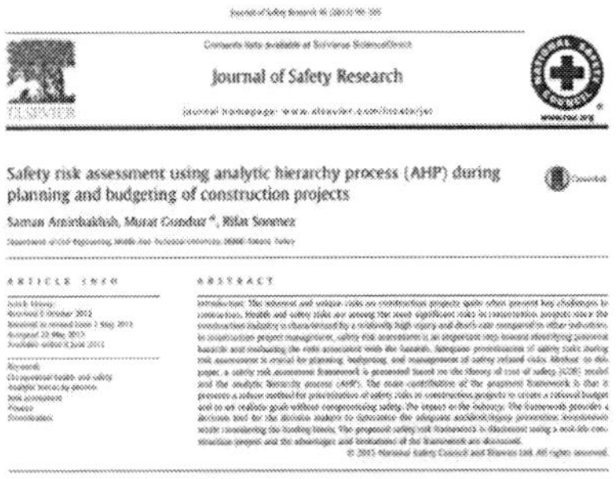

IV. Safety indexes design

- Preferences elicitation through a set of organized interactions with the EPSF personnel.
- Approach.
  1. Elaboration of the Individual preferences questionnaire and collection of answer.
  2. Analysis and preparation of the workshop
  3. Collective workshop.
  4. Uncertainties management and validation of weights.

IV. Safety indexes design

- **Approach**
  - Comparison of the intensity of preferences between each pairs of levels.
  - For instance: "to which extent the difference between L1 and L2 incidents is higher, equal or lower than between L2 and L3?

1	Minor safety event
2	Event that may have led to material consequences or minor wounds
3	Event that may have led to serious human consequences (1 to 2 serious injuries – 24h hospitalization) or one fatality.
4	Event that may have led to large human consequences (several major injuries and fatalities)

  - Individual appreciations of these differences revealed different from one participant to another.
  - The workshop discussed these preferences and led to three different configurations to be tested

INERIS
maîtriser le risque | pour un développement durable |

15

IV. Safety indexes design

- **Results**
  - W_1: 5%
  - W_2: 9%
  - W_3: 24%
  - W_4: 62%
  - W_5: 25%
  - W_6: 75%

INERIS
maîtriser le risque | pour un développement durable |

16

IV. Safety indexes design

- **Validation**

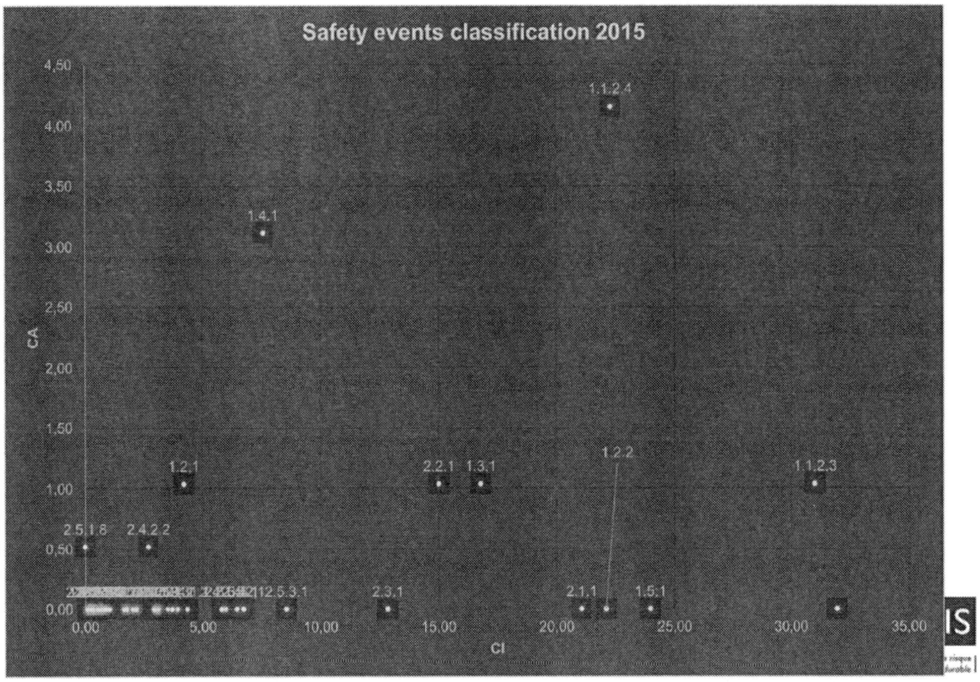

IV. Safety indexes design

- **Validation**

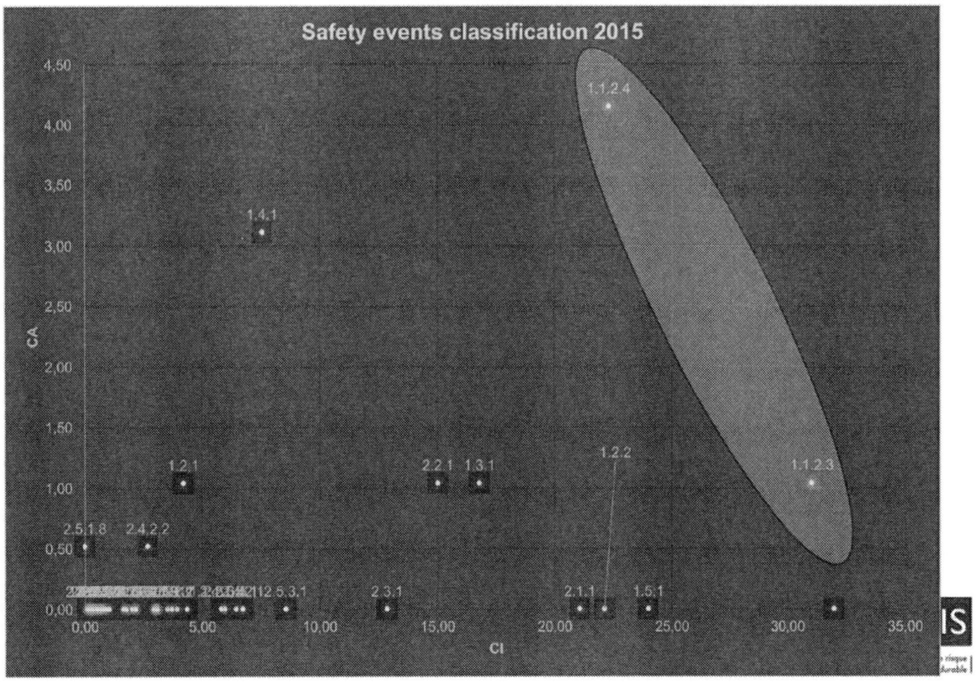

IV. Safety indexes design

- **Validation**

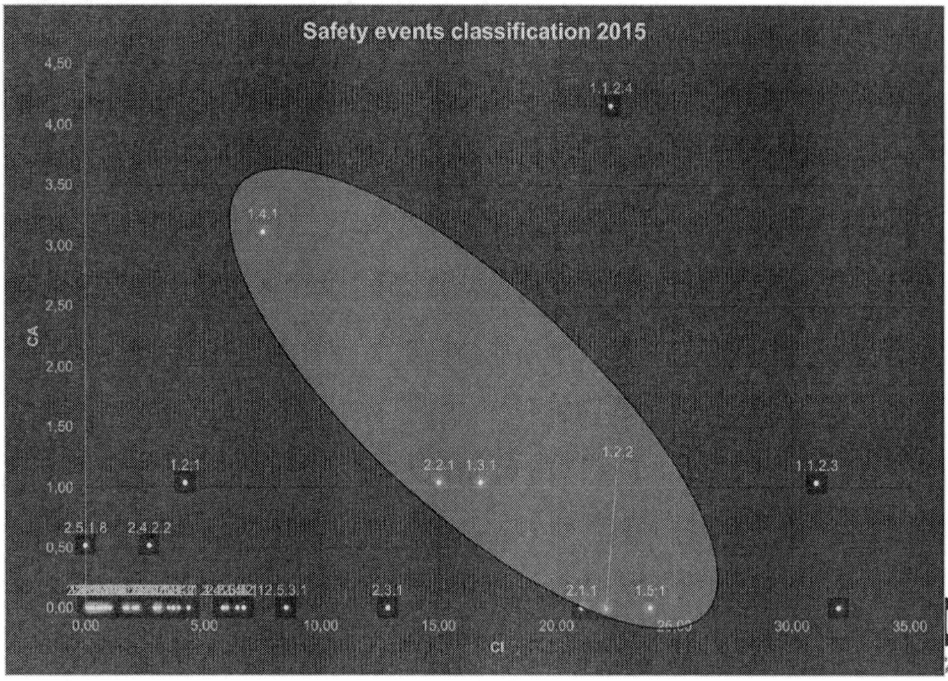

IV. Validation des indices

- **GSL**

YEARS	AC_G	IC_G	Comments
2015	0,01	13,34	• First year.
2016	0,10	20,19	• Increase in current final risk levels and negative perspectives given the rise of incidental records. • Important variations in incident reporting with a possible impact on data quality.
2017	4,42	14,97	• The incidents increase pointed last year seems to result in higher accidental records. • With reporting capabilities being unchanged, a decrease in incidental records appears as encouraging. However, further analysis should discuss whether this decrease can be linked to specific and identifiable courses of actions.

INERIS

maîtriser le risque | pour un développement durable |

20

V. Recommandations

- Do not delegate decisions to indicators
  - The suggested indexes are not an objective or exhaustive description of reality.
  - They are meant to be good starting points for an enhanced collective thinking on safety matters.
- Articulation with other EPSF control levers
  - Suggested indexes are nothing but an additional lens through which railway operators can be assessed.
- Work to come
  - Homogenization of reporting practices.
  - Relate events reported to their causes and associated SMS processes.

INERIS
maîtriser le risque | pour un développement durable |

21

2019 EUROPEAN CONFERENCE ON PROCESS SAFETY & BIG DATA

CCPS EPSC

PETRONEW engenharia industrial

UNICAMP

FACULDADE DE ENGENHARIA QUÍMICA UNICAMP

Using an Artificial Intelligence Technique and Quantitative Data in Risk Assessment Matrix: Fuzzy Sets Approach

Pedro Prado - Author

Prof. Dr. Adilson Afonso - Co-Author

Prof. Dr. Leonardo Fregolente - Co-Author

Frankfurt am Main, Germany - October/2019

Agenda

- Introduction and Justification
- Conventional Risk Matrix
- Fuzzy Method
  - What is Fuzzy Theory?
  - Fuzzification
  - If-Then Rules
  - Inference Machine
  - Deffuzification
- Results: Fuzzy Risk Matrix
- Conclusion
- References

Introduction and Justification

- Risk assessment: provides a basis for proposing measures to reduce and mitigate scenarios that can cause harm to people, the environment, and property.

In industry:

Risk = Frequency of Occurrence x Severity of the Consequence

- A common practice is to define industrial risk in a qualitative way.

Introduction and Justification

- Purely qualitative:
  - It can lead to imprecision and uncertainties.
  - Risk could be underestimated or overestimated in the process safety evaluation.
- Companies are trying to reduce and quantify uncertainties in Risk Assesements.
- Quantitative data can reduce signicantly the uncertainties.

Introduction and Justification

- Fuzzy Logic is a rule-based system that can rely on the practical experience and manipulates mathematically linguistic concepts that may be vague and imprecise.

- In this work, we use fuzzy theory to mathematically manipulate the linguistic terms of severity and frequency in risk level calculation.

Conventional Risk Matrix

- Conventional Risk Matrix is usually based in Frequency and Severity categories.

Risk Categories	Description
Tolerable (T) LEVEL 1	There is no need for additional measures. Monitoring is necessary to ensure that controls are maintained.
Moderate (M) LEVEL 2	Additional measures should be evaluated with a view to achieving risk reduction and implementing those considered feasible (ALARP - As Low As Reasonably Practicable).
Not Tolerable (NT) LEVEL 3	Existing Controls are insufficient. Alternative methods should be considered to reduce the likelihood of occurrence or severity of consequences to bring risk to regions of lower risk magnitude (ALARP or Tolerable regions).

Fuzzy Method: What is Fuzzy Theory?

- Fuzzy logic is a set of mathematical principles for knowledge representation based on degrees of membership.
- It considers approximations or subjective numbers and quantifies the ambiguity and vagueness of linguistic descriptions.
- Unlike two-valued Boolean logic, fuzzy logic is multi-valued. It deals with degrees of membership and degrees of truth.

Fuzzy Method: What is Fuzzy Theory?

- Fuzzy logic uses the continuum of logical values between 0 (completely false) and 1 (completely true).
- Instead of just black and white, it employs the spectrum of colors, accepting that things can be partly true and partly false at the same ti

Fuzzy Method: Fuzzification

- Fuzzification is the process of changing a real scalar value into a fuzzy value (membership degree)
- This transformation is made by membership functions.

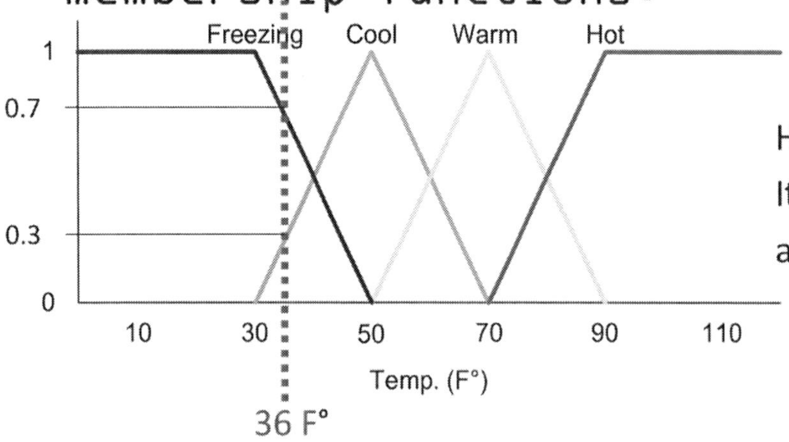

How cool is 36 F° ?

It is 30% Cool

and 70% Freezing

Fuzzy Method: Fuzzification

- We define the boundaries of the crisp sets of frequency, severity and risk.

			Frequency Category				
			10^{-5}-10^{-4}	10^{-4}-10^{-3}	10^{-3}-10^{-2}	10^{-2}-10^{-1}	10^{-1}-1
			Extremely remote	Remote	Unlikely	Likely	Frequent
Severity Category	4,5 - 5,5	Catastrophic	Moderate (M)	Moderate (M)	Not Tolerable (NT)	Not Tolerable (NT)	Not Tolerable (NT)
	3,5 - 4,5	Critical	Tolerable (T)	Moderate (M)	Moderate (M)	Not Tolerable (NT)	Not Tolerable (NT)
	2,5 - 3,5	Medium	Tolerable (T)	Tolerable (T)	Moderate (M)	Moderate (M)	Not Tolerable (NT)
	1,5 - 2,5	Marginal	Tolerable (T)	Tolerable (T)	Tolerable (T)	Moderate (M)	Moderate (M)
	0,5 - 1,5	Negligible	Tolerable (T)	Tolerable (T)	Tolerable (T)	Tolerable (T)	Moderate (M)

Fuzzy Method: Fuzzification

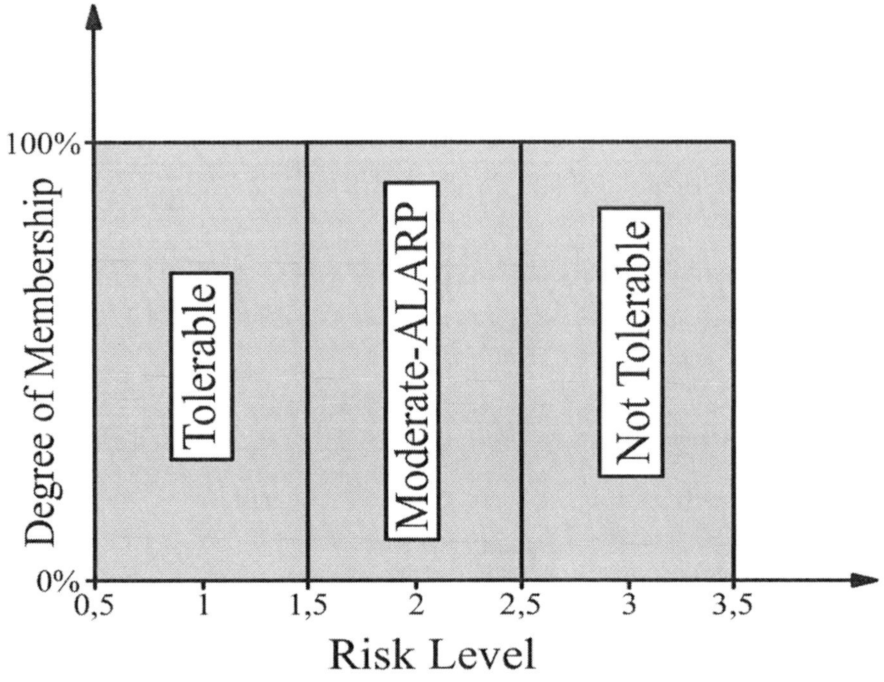

Fuzzy Method: Fuzzification

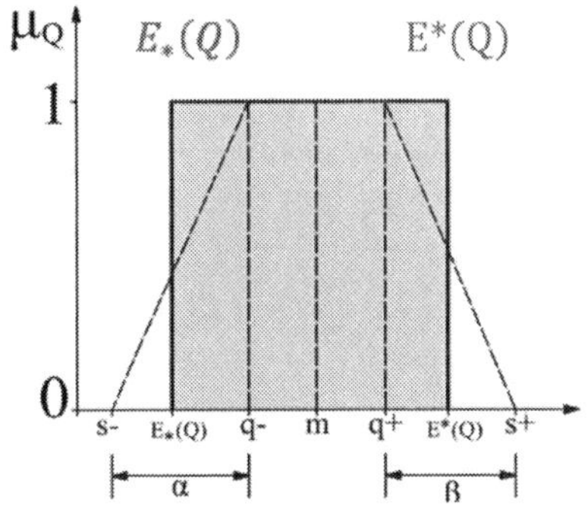

$$\mu_Q(x) = \begin{cases} 0, & x < s_- \\ \dfrac{x - s_-}{q_- - s_-}, & s_- \leq x \leq q_- \\ 1, & q_- \leq x \leq q_+ \\ \dfrac{s_+ - x}{s_+ - q_+}, & q_+ \leq x \leq s_+ \\ 0, & s_+ < x \end{cases}$$

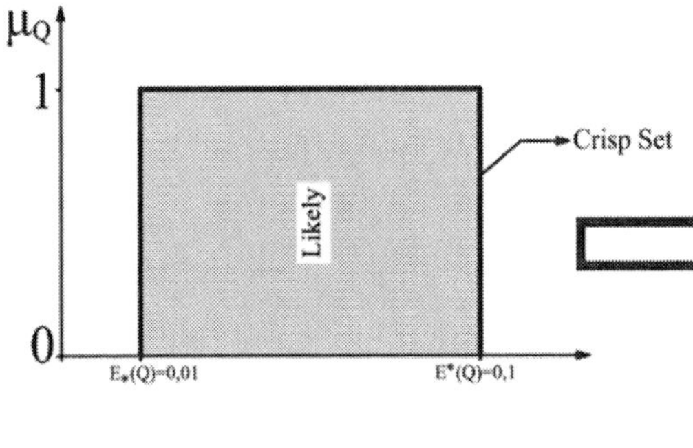

Trapezoidal Membership Function

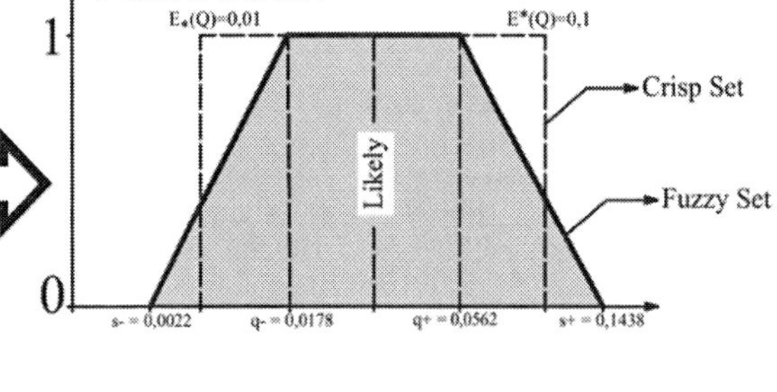

Fuzzy Method: Fuzzification

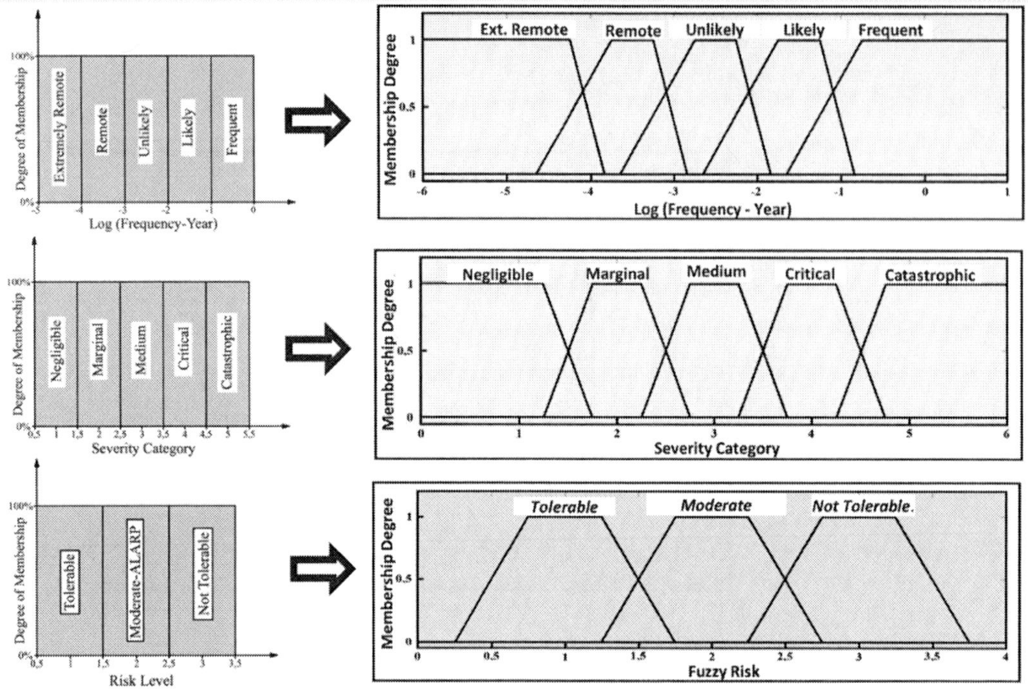

Fuzzy Method: If-Then Rules

- Rules If-Then: incorporation of human experience.

N°	Fuzzy IF- THEN Rules
#1	If (Freq.) is ***Frequent*** AND (Sev.) is ***Negligible*** Then (Fuzzy-Risk) is ***Moderate***.
#2	If (Freq.) is ***Frequent*** AND (Sev.) is ***Marginal*** Then (Risco-Fuzzy) is ***Moderate***.
#3	If (Freq.) is ***Frequent*** AND (Sev.) is ***Medium*** Then (Risco-Fuzzy) is ***Not Tolerable***.
#4	If (Freq.) is ***Frequent*** AND (Sev.) is ***Critical*** Then (Risco-Fuzzy) is ***Not Tolerable***.
#5	If (Freq.) is ***Frequent*** AND (Sev.) is ***Catastrophic*** Then (Risco-Fuzzy) is ***Not Tolerable***.
#6	If (Freq.) is ***Likely*** AND (Sev.) is ***Negligible*** Then (Risco-Fuzzy) is ***Tolerable***.
#7	If (Freq.) is ***Likely*** AND (Sev.) is ***Marginal*** Then (Risco-Fuzzy) is ***Moderate***.
#8	If (Freq.) is ***Likely*** AND (Sev.) is ***Medium*** Then (Risco-Fuzzy) is ***Moderate***.
#9	If (Freq.) is ***Likely*** AND (Sev.) is ***Critical*** Then (Risco-Fuzzy) is ***Not Tolerable***.
#10	If (Freq.) is ***Likely*** AND (Sev.) is ***Catastrophic*** Then (Risco-Fuzzy) is ***Not Tolerable***.
#11	If (Freq.) is ***Unlikely*** AND (Sev.) is ***Negligible*** Then (Risco-Fuzzy) is ***Tolerable***.
#12	If (Freq.) is ***Unlikely*** AND (Sev.) is ***Marginal*** ThenThen (Risco-Fuzzy) is ***Tolerable***.
#13	If (Freq.) is ***Unlikely*** AND (Sev.) is ***Medium*** Then (Risco-Fuzzy) is ***Moderate***.
#14	If (Freq.) is ***Unlikely*** AND (Sev.) is ***Critical*** Then (Risco-Fuzzy) is ***Moderate***.
#15	If (Freq.) is ***Unlikely*** AND (Sev.) is ***Catastrophic*** Then (Risco-Fuzzy) is ***Not Tolerable***.
#16	If (Freq.) is ***Remote*** AND (Sev.) is ***Negligible*** Then (Risco-Fuzzy) is ***Tolerable***.
#17	If (Freq.) is ***Remote*** AND (Sev.) is ***Marginal*** Then (Risco-Fuzzy) is ***Tolerable***.
#18	If (Freq.) is ***Remote*** AND (Sev.) is ***Medium*** Then (Risco-Fuzzy) is ***Tolerable***.
#19	If (Freq.) is ***Remote*** AND (Sev.) is ***Critical*** Then (Risco-Fuzzy) is ***Moderate***.
#20	If (Freq.) is ***Remote*** AND (Sev.) is ***Catastrophic*** Then (Risco-Fuzzy) is ***Moderate***.
#21	If (Freq.) is ***Ext. Remote*** AND (Sev.) is ***Negligible*** Then (Risco-Fuzzy) is ***Tolerable***.
#22	If (Freq.) is ***Ext. Remote*** AND (Sev.) is ***Marginal*** Then (Risco-Fuzzy) is ***Tolerable***.
#23	If (Freq.) is ***Ext. Remote*** AND (Sev.) is ***Medium*** Then (Risco-Fuzzy) is ***Tolerable***.
#24	If (Freq.) is ***Ext. Remote*** AND (Sev.) is ***Critical*** Then (Risco-Fuzzy) is ***Tolerable***.

- Rules were based in conventinal risk matrix.

Severity Category			Frequency Category				
					III	IV	V
			Extremely remote	Remote	Unlikely	Likely	Frequent
	V	Catastrophic	Moderate (M)	Moderate (M)	Not Tolerable (NT)	Not Tolerable (NT)	Not Tolerable (NT)
	IV	Critical	Tolerable (T)	Moderate (M)	Moderate (M)	Not Tolerable (NT)	Not Tolerable (NT)
	III	Medium	Tolerable (T)	Tolerable (T)	Moderate (M)	Moderate (M)	Not Tolerable (NT)
	II	Marginal	Tolerable (T)	Tolerable (T)	Tolerable (T)	Moderate (M)	Moderate (M)
	I	Negligible	Tolerable (T)	Tolerable (T)	Tolerable (T)	Tolerable (T)	Moderate (M)

Fuzzy Method: Inference Machine

- Event: Frequency: 95 years; Severity: II (Marginal)

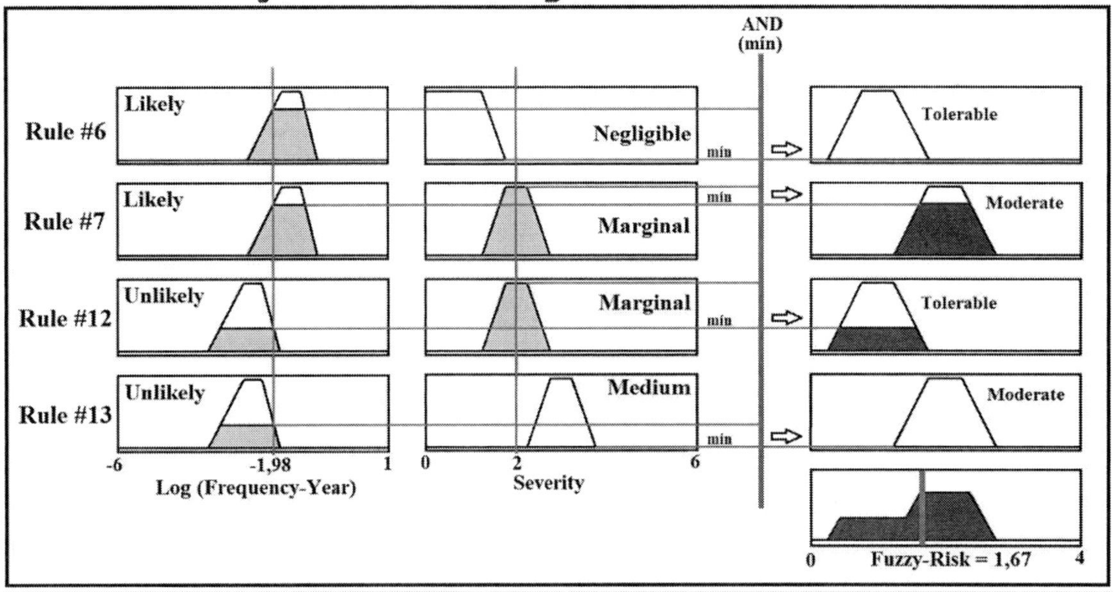

Fuzzy Method: Deffuzification

- Defuzzification: obtaining a single number from the output of the aggregated fuzzy set.
- Centroid defuzzification method finds a point representing the centre of gravity of the fuzzy set

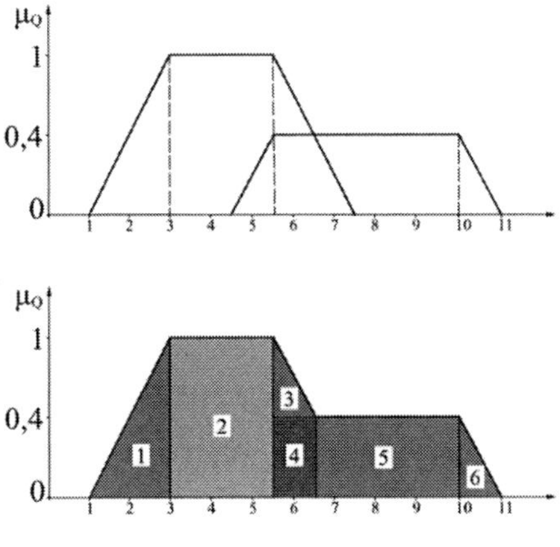

$$X^* = \frac{\sum_{i=1}^{N} A_i \times \overline{X}_i}{\sum_{i=1}^{N} A_i}$$

Where **N** indicates the number of subareas and A_i and $\overline{X_i}$ represent, respectively, the area and center area of each subarea i^{th}.

Results: Fuzzy Risk

- A new Risk Matrix was gerated.

- The input of this matriz are:
  - Log (Frequency in Years)
  - Severity

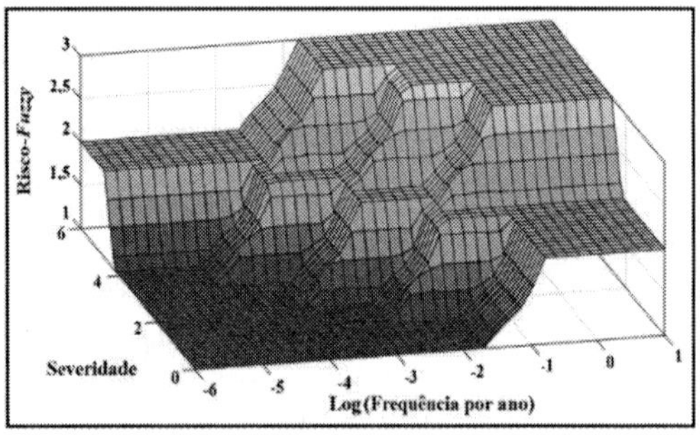

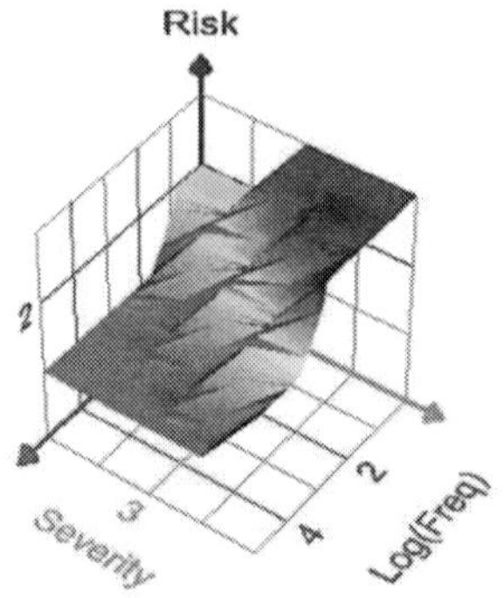

Results: Fuzzy Risk

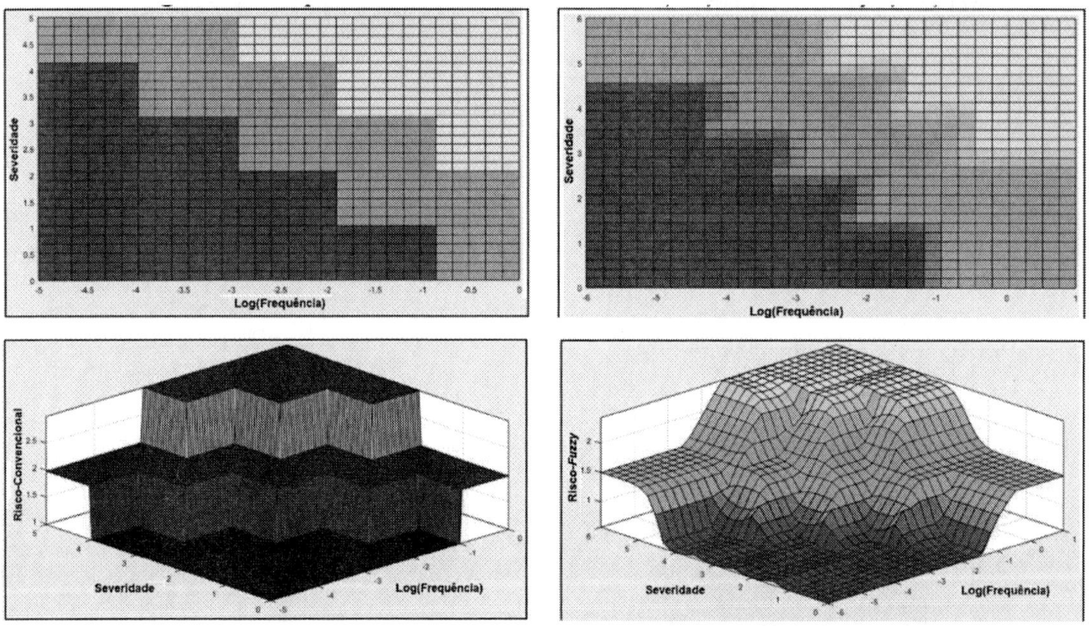

Results: Fuzzy Risk

- Fuzzy-Risk in Microsoft Excel Aplication.
- VBA coding.
- We integrated it in a Hazop

Deviations	Causes	Safeguards	Consequences	Frequency	Severity	Fuzzy-Risk	Recommendations
High Pressure	Control Valve Failure (Fail Open)	Pressure Relief Valve (PSV)	Pipe Rupture, large leak with explosion	III	V	III	Redundant safety valve installation to reduce the likelihood of the event occurring.
Low Pressure	Shut Down / Pump Failure	Pressure Transmitter (PIT)	Plant Shut Down causing financial losses	II	II	II	Spare pump installation to keep unit running.

⇩

Deviations	Causes	Safeguards	Consequences	Frequency (Years)	Severity	Fuzzy-Risk	Recommendations
High Pressure	Control Valve Failure (Fail Open)	Pressure Relief Valve (PSV)	Pipe Rupture, large leak with explosion	950	5	2,67	Redundant safety valve installation to reduce the likelihood of the event occurring.
Low Pressure	Shut Down / Pump Failure	Pressure Transmitter (PIT)	Plant Shut Down causing financial losses	97	2	1,65	Spare pump installation to keep unit running.

Results: Fuzzy Risk

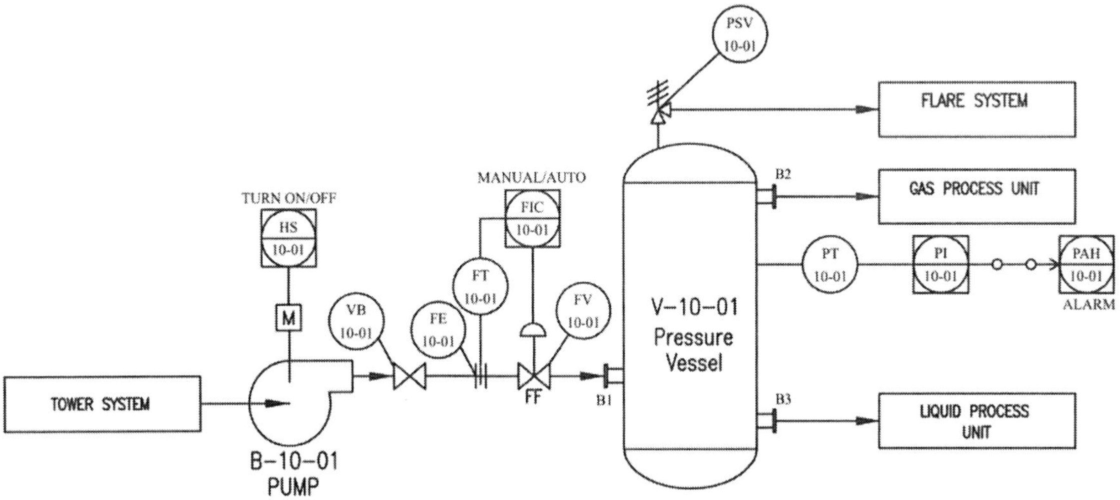

Results: Fuzzy Risk

HAZOP SPREADSHEET

SYSTEM: Transfer to gas / liquid separation vessel
NODE: DA ENTRADA DA BOMBA DE EXPEDIÇÃO B-01 ATÉ O BOCAL DE ENTRADA DE PRODUTO B3

PARTICIPANT: Bruno (Processo), Adilson, Claudinei Broio (Mec), Elmer Dennis (Eletrica), Pedro (Instrumentação)

Item	Deviation	Cause	Safeguards	Consequence	Frequency	Severity	Risk (Conventional)	Frequency (Years)	Severity	Fuzzy-Risk
1	HIGH PRESSURE	Control valve failure FCV-10-01 failing closed. (Due to flow transmitter failure, logic or valve failure)	Flow Measurement (FT-10-01) is considered a safeguard for the zero flow scenario as the zero flow is easily perceived by the operation team.	Small product leakage in pipe flanges.	5	2	2	80	2	1,83
2	HIGH PRESSURE	Control valve failure FCV-10-01 failing closed. (Due to flow transmitter failure, logic or valve failure)	Flow Measurement (FT-10-01) is considered a safeguard for the zero flow scenario as the zero flow is easily perceived by the operation team.	#NOME?	2	5	2	33333	5	2,00
3	HIGH PRESSURE	Control valve failure FCV-10-01 failing closed. (Due to flow transmitter failure, logic or valve failure)	Flow Measurement (FT-10-01) is considered a safeguard for the zero flow scenario as the zero flow is easily perceived by the operation team.	Plant shutdown with production returning in hours.	5	2	2	10	2	2,00
4	HIGH PRESSURE	Control valve failure FCV-10-01 failing closed. (Due to flow transmitter failure, logic or valve failure)	Flow Measurement (FT-10-01) is considered a safeguard for the zero flow scenario as the zero flow is easily perceived by the operation team.	Temperature increse in pump B-10-01 causing damage to equipment (zero flow). Repair will occur in days.	5	3	3	20	3	2,32
5	HIGH PRESSURE	Control valve failure FCV-10-01 failing closed. (Due to flow transmitter failure, logic or valve failure)	Flow Measurement (FT-10-01) is considered a safeguard for the zero flow scenario as the zero flow is easily perceived by the operation team.	Excessive vibration of the B-10-01 pump may cause misalignment and bearing problems.	5	1	2	20	1	1,32
6	HIGH PRESSURE	FCV-10-01 Control Valve Fault with Operator Commanded Improper Shutdown (PID Controller in Manual)	Flow Measurement (FT-10-01) is considered a safeguard for the zero flow scenario as the zero flow is easily perceived by the operation team.	Small product leakage in pipe flanges.	5	2	2	40	2	2,00

Conclusion

- The risk obtained with Fuzzy set theory proved to be more realistic and precise since Fuzzy Risk presents a gradual and continuous transition between the categories of the risk matrix.

- With the aid of quantitative data, the risks were, in general, smaller. Hazop-Fuzzy allows easy incorporation of actual occurrence rate data for a particular hazardous event.

References

CENTER FOR CHEMICAL PROCESS SAFETY OF THE AMERICAN INSTITUTE OF CHEMICAL ENGINEERS. **Guidelines for Safe and Reliable Instrumented Protective Systems.** John Wiley & Sons, Inc., Chichester-UK, 2007. 430 p.

CENTER FOR CHEMICAL PROCESS SAFETY OF THE AMERICAN INSTITUTE OF CHEMICAL ENGINEERS. **Layer of Protection Analysis Simplified Process Risk Assessment**. CCPS, New York, 2001. 270 p.

DUBOIS, D.; PRADE, H. The mean value of a fuzzy number. **Fuzzy Sets and Systems**. v. 24, p. 279–300, 1987.

ERICSON, C. A. **Hazard Analysis Techniques for System Safety**. 2. Ed., John Wiley & Sons, Inc., Hoboken-NJ, 2015. 640 p.

HEALTH & SAFETY EXECUTIVE. **Reducing risks, protecting people: HSE's decision-making process.** HSE Books, Norwich-UK, 2001. 88 p.

HEALTH & SAFETY EXECUTIVE. **The Tolerability of Risk from Nuclear Power Stations.** HSE Books, London-UK, 1992. 65 p.

KAPLAN, S.; GARRICK, B, J. On the qualitative definition of risk. **Risk Analysis**, v. 1, p. 11-27, 1981.

KING, R. **Safety in the Process Industries**. Butterworth-Heinemann, 1990. 780 p.

KLETZ, T. A. **Hazop & Hazan: Identifying and Assessing Process Industry Hazards**. 4. Ed, IchemE, 1999. 232 p.

KLETZ, T. A. **What Went Wrong?**. 5. Ed, Butterworth-Heinemann; 2009. 640 p.

LEES, F. **Lees' Loss prevention in the process industries: Hazard identification, assessment and control**. 4. Ed.,.

MANNAN, M. S. **Lees' Process Safety Essentials: Hazard Identification, Assessment and Control**. Butterworth-Heinemann, 2013. 570 p.

2019 EUROPEAN CONFERENCE ON PROCESS SAFETY & BIG DATA

CCPS EPSC

PETRONEW engenharia industrial

UNICAMP

FACULDADE DE ENGENHARIA QUÍMICA UNICAMP

Pedro Prado
Engineering and Contract Manager |
MSc Chemical Engineering | ISA/IEC 61...

Thank You!

Pedro Prado - Author
Prof. Dr. Adilson Afonso - Co-Author
of. Dr. Leonardo Fregolente - Co-Author

enablon

Integrating Operational And Enterprise Risk Management To Improve Asset Integrity And Reliability

Presenters: Igor Milic, Sam Williams
Contributor: Glenn Pettitt

ERM

MOLGROUP

This is us

High temperatures, high pressures, reactive substances, everything is flammable and explosive. Safety is highly dependent on asset integrity. That is why we are conducting process hazard analyses. We try to recognize and then manage as many risks as possible.

Our Operational Risk*

01 Tens of thousands of recognized risks
PHA studies generate many possible scenarios and each scenario can have multiple recommendations.

02 Risks appear constantly
No matter how many studies are conducted, risks do not disappear and they need to be managed.

03 A lot of expertise is needed
After the risk is recognized, it requires some serious technical knowledge to assess it.

04 Resources
Each risk requires some resources (money or people) to be managed. The resources are not limitless and priorities need to be set.

BOTTOM UP
The process is fed bottom up. From recognizing, through assessing to prioritizing - there is always a possibility to process wrong piece of information. The consequences can be deadly. Literally.

*Although Operational risk has different connotations for different businesses, we at MOL Group have decided to use the phrase as an umbrella term for managing process safety risks.

LIKE ALL BIG COMPANIES, WE ARE ALSO CONCERNED WITH ENTERPRISE RISK MANAGEMENT

AS YOU CAN IMAGINE, THERE ARE MANY HIGH-LEVEL ASPECTS IMPACTING HIGH-LEVEL DECISIONS

So, We Started Thinking

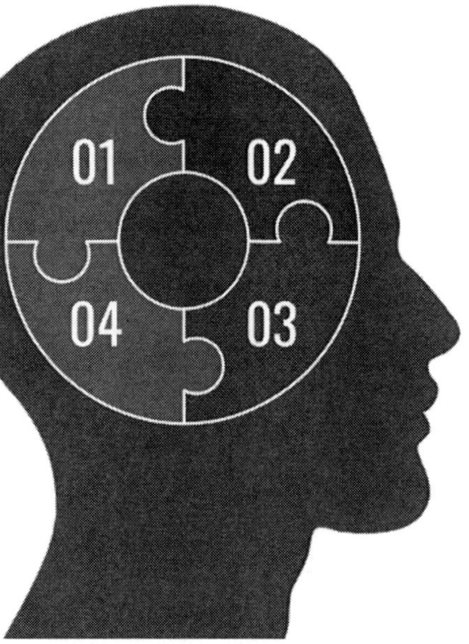

Operational risk needs a strong system to be managed. It should be flexible enough to adapt to our everchanging needs. The system needs to accomodate the technical assessment of risks. And to provide some meaningful insights for budgeting decisions.

But, wait..

What if, at the same time, **we could integrate Operational and Enterprise risk** management?

01
Tens of thousands of recognized risks

02
Risks appear constantly

03
A lot of expertise is needed

04
Resources

And We Came Up With This Process

(01) Risk Input
Risk Input is an umbrella record which captures all recommendations from a single PHA study (1 Risk Input = 1 Study).

(02) Specific Risk
Each recommendation from the PHA study is connected to a certain risk. The risk and the recommendation are captured as a Specific Risk record.

(03) Expertise Review
Each Specific Risk is reviewed by a team of relevant experts. They decide whether to accept or reduce it. They then recognize the Operational Risk category Specific Risk is tied to.

Action Plans (04)
If the Specific Risk is to be reduced, the team of experts creates the Action Plan for the risk reduction measures (prevention, control and/or mitigation). And they assess if the budget approval is needed.

Operational Risk (05)
Based on the Specific Risks information (at the Unit level) and their categories, Process Safety expert reviews each category of the Operational Risk at Unit Block level.

Risk Aggregation (06)
Similar to the Specific Risk being aggregated from Unit level to Operational Risk categories at the Unit Block level, Operational Risks are assessed further at the Site and Group levels.

The creation of

Building Blocks

Using the agile approach, we developed the build-up from the unit level Specific risk to Site/Group level Operational Risk. The entire process fully utilized Enablon platform and respected the MOL Group structure.

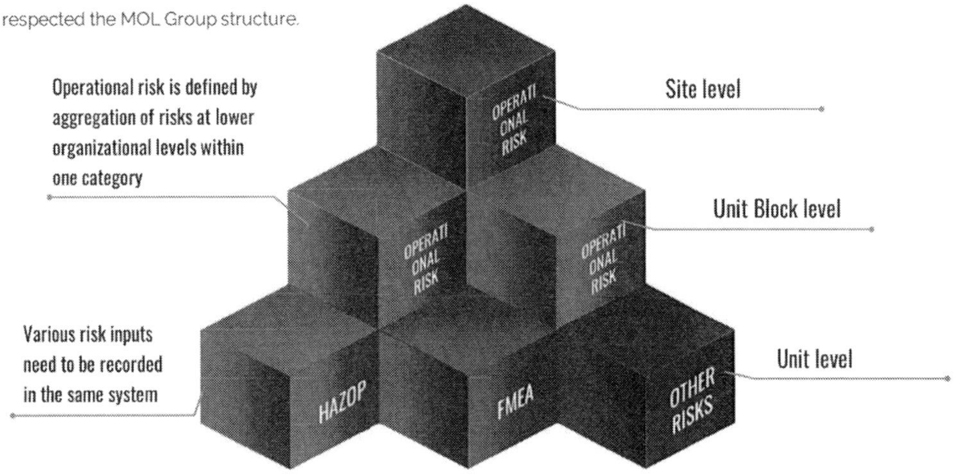

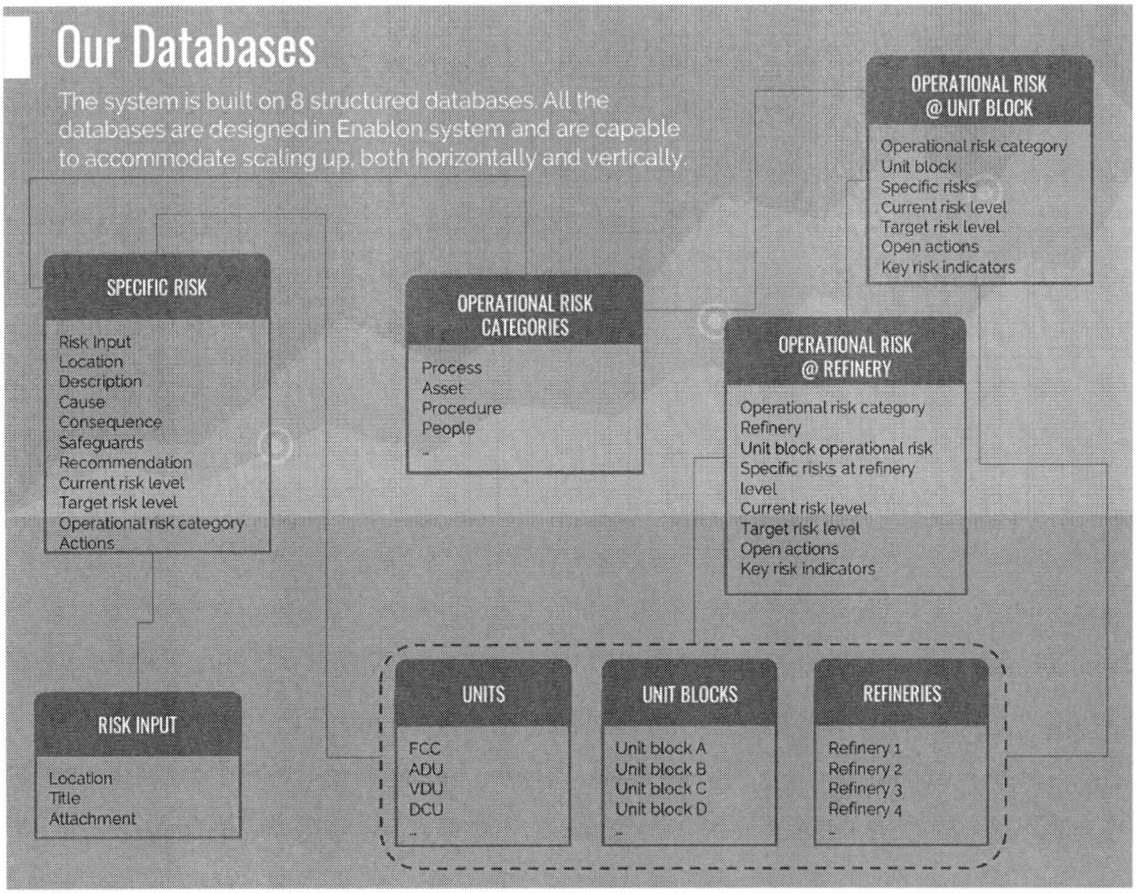

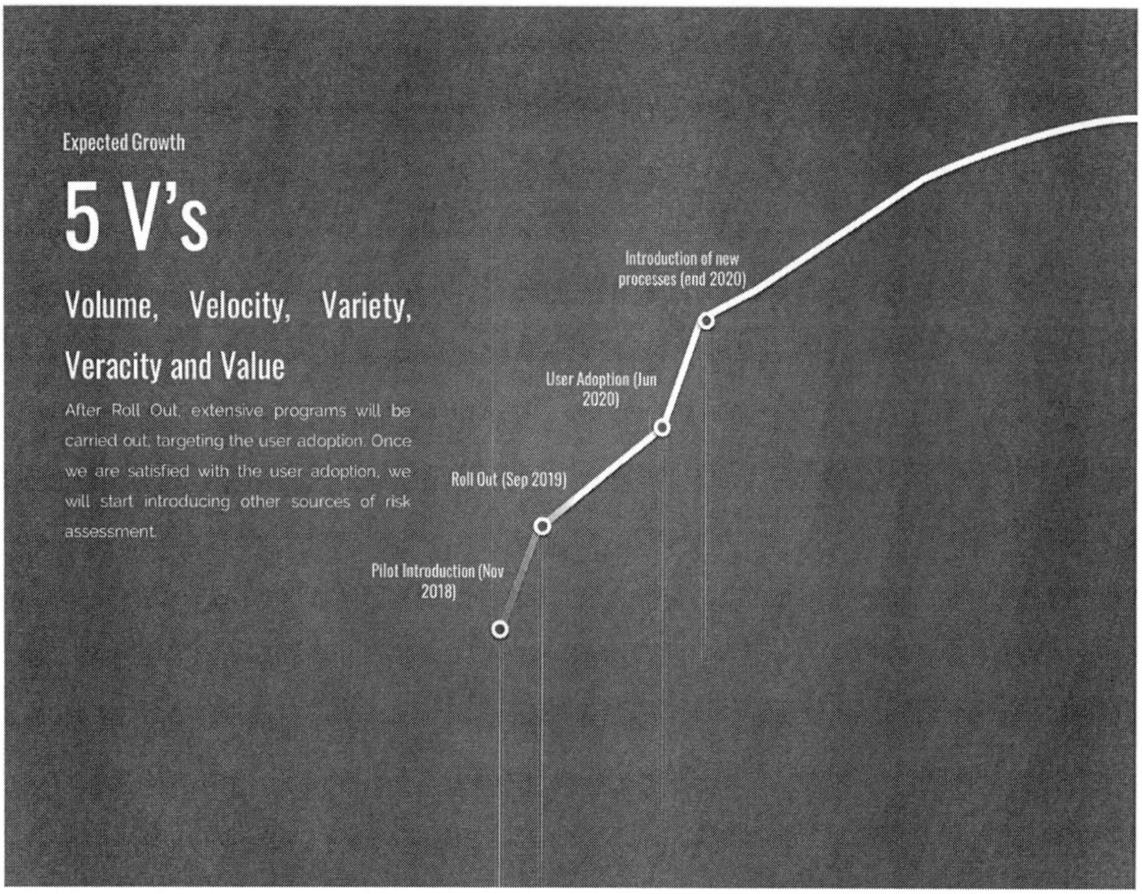
Expected Growth
5 V's
Volume, Velocity, Variety, Veracity and Value
After Roll Out, extensive programs will be carried out, targeting the user adoption. Once we are satisfied with the user adoption, we will start introducing other sources of risk assessment.
Introduction of new processes (end 2020)
User Adoption (Jun 2020)
Roll Out (Sep 2019)
Pilot Introduction (Nov 2018)

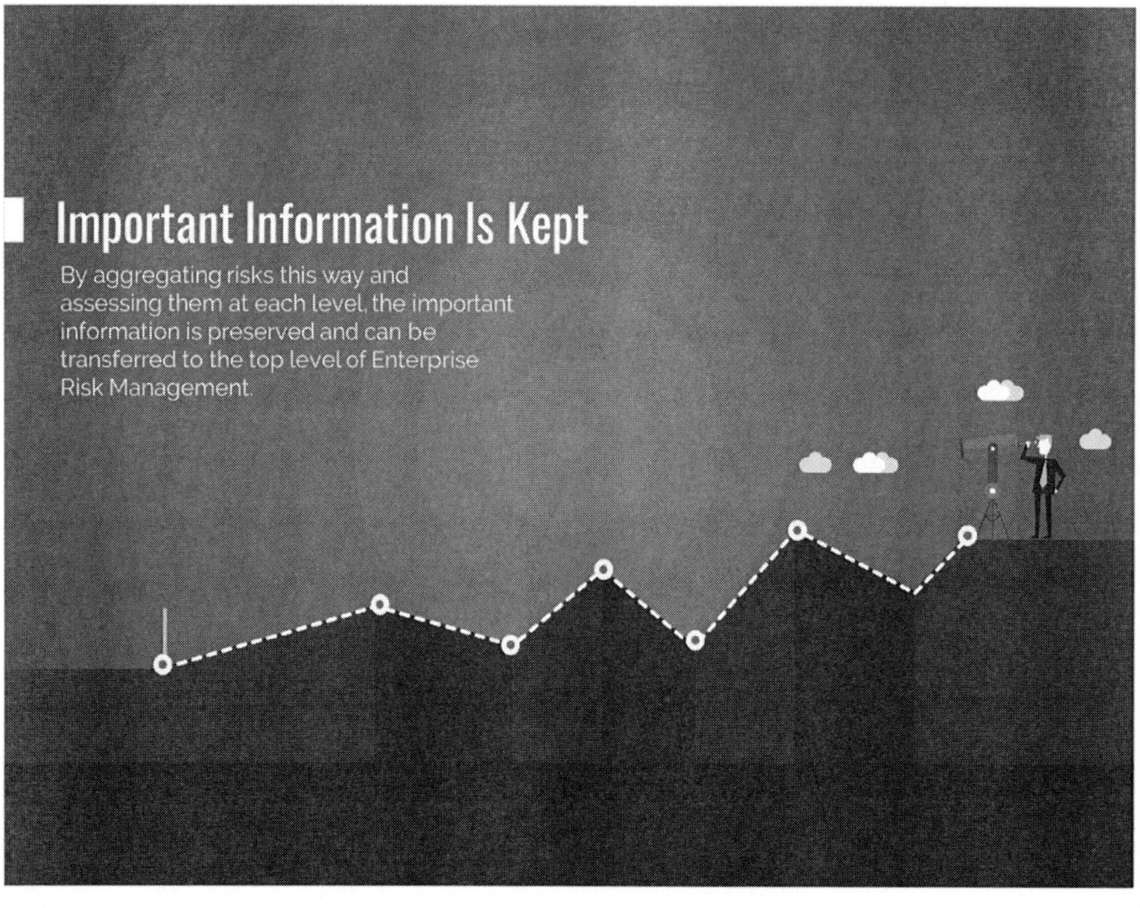
Important Information Is Kept
By aggregating risks this way and assessing them at each level, the important information is preserved and can be transferred to the top level of Enterprise Risk Management.

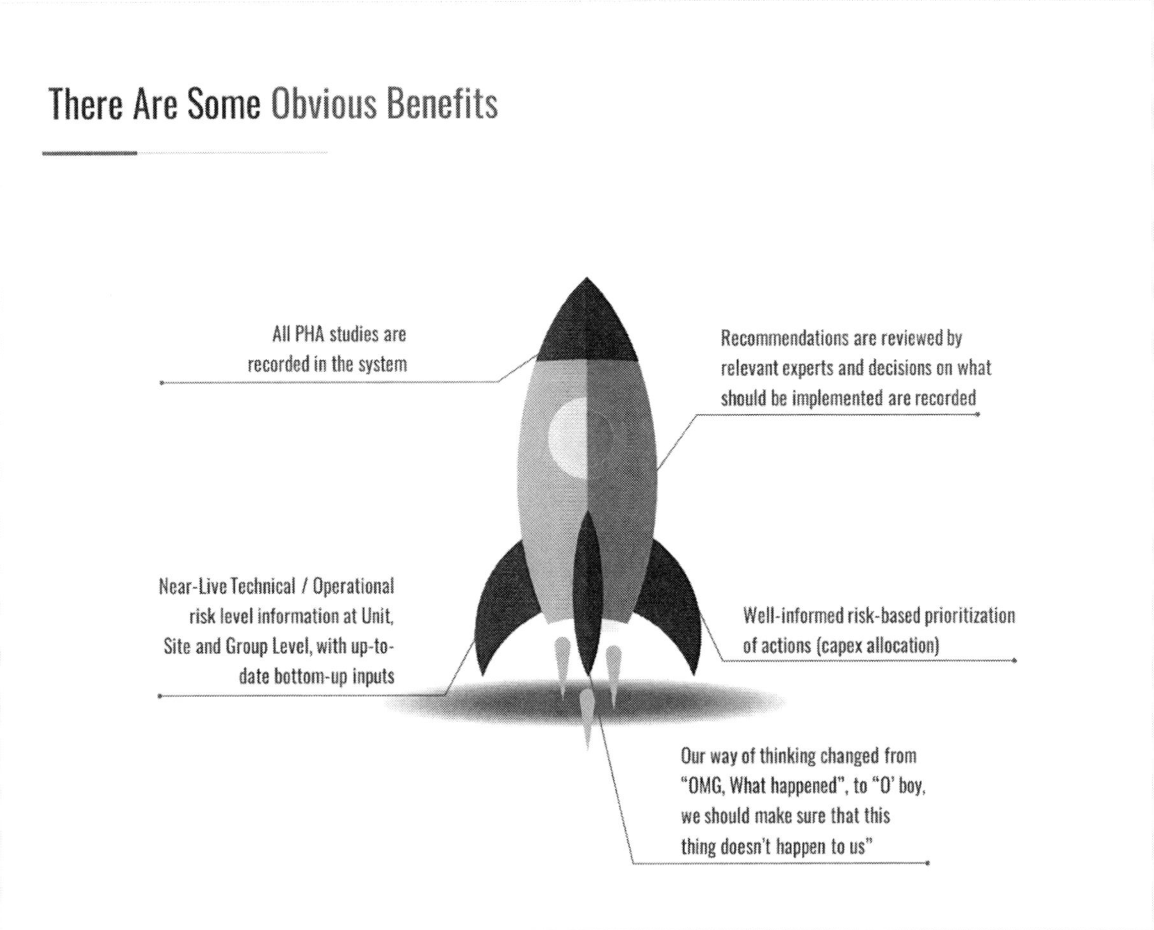
There Are Some Obvious Benefits
All PHA studies are recorded in the system
Recommendations are reviewed by relevant experts and decisions on what should be implemented are recorded
Near-Live Technical / Operational risk level information at Unit, Site and Group Level, with up-to-date bottom-up inputs
Well-informed risk-based prioritization of actions (capex allocation)
Our way of thinking changed from "OMG, What happened", to "O' boy, we should make sure that this thing doesn't happen to us"

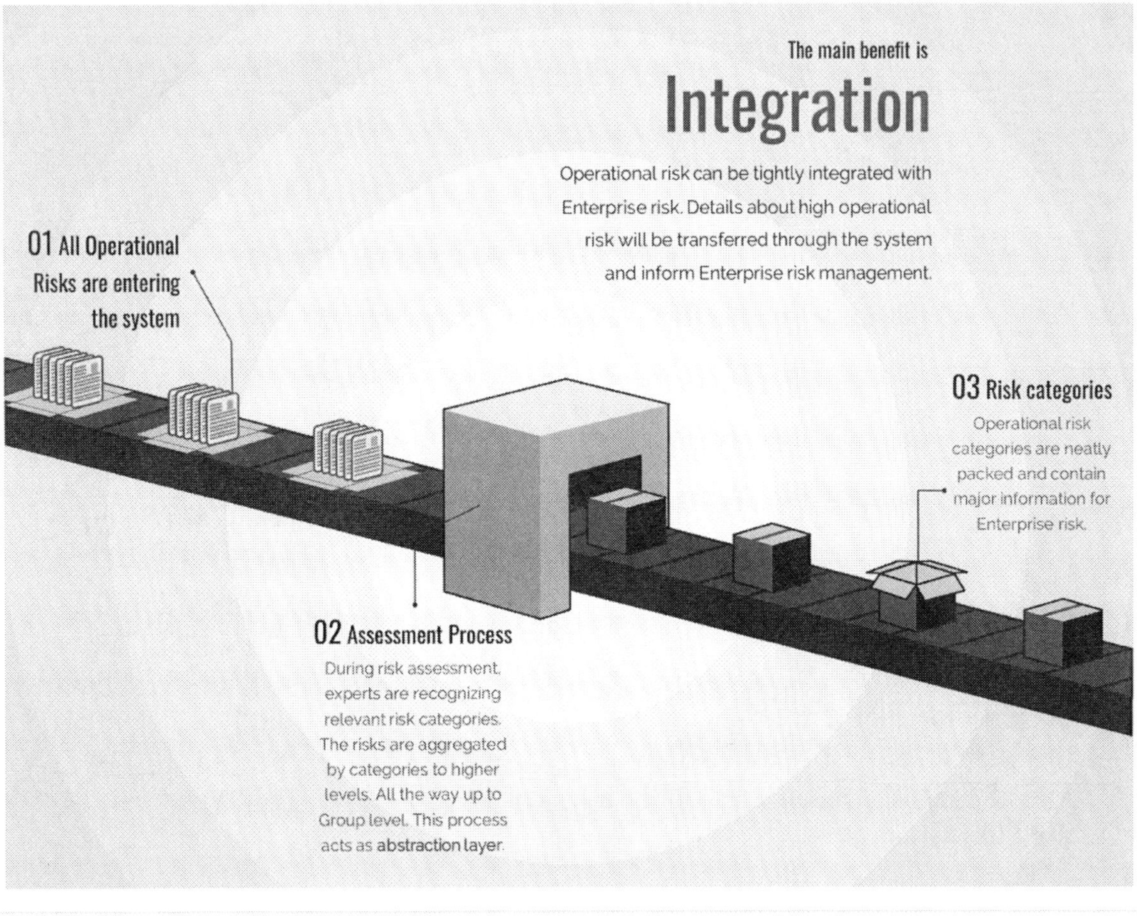
The main benefit is
Integration
Operational risk can be tightly integrated with Enterprise risk. Details about high operational risk will be transferred through the system and inform Enterprise risk management.
01 All Operational Risks are entering the system
02 Assessment Process
During risk assessment, experts are recognizing relevant risk categories. The risks are aggregated by categories to higher levels. All the way up to Group level. This process acts as abstraction layer.
03 Risk categories
Operational risk categories are neatly packed and contain major information for Enterprise risk.

WE BROUGHT DOWN THE WALL BETWEEN ENTERPRISE RISK AND OPERATIONAL RISK MANAGEMENT

WE HAVE TURNED OPERATIONAL RISK INFORMATION INTO VALUABLE ENTERPRISE RISK INSIGHTS

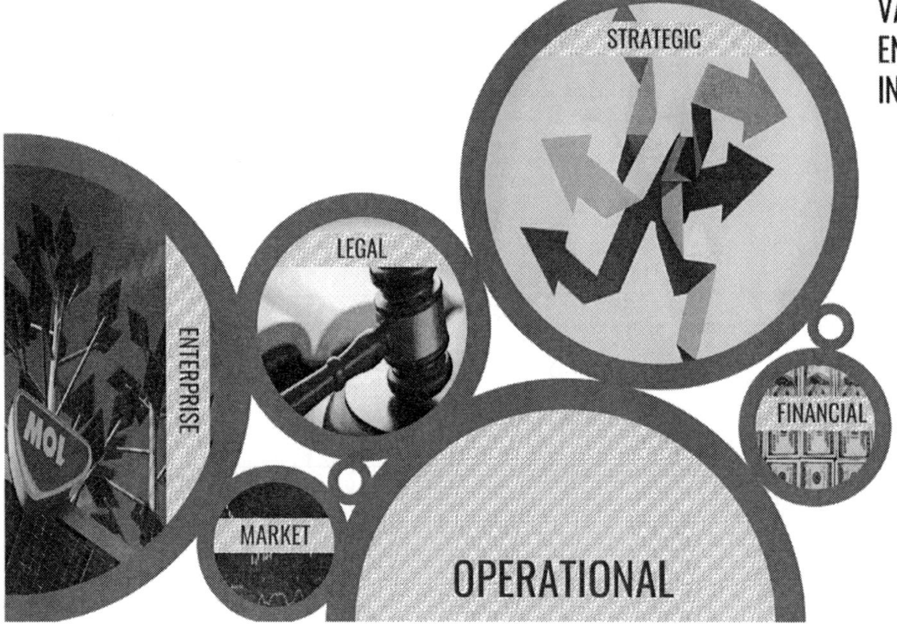

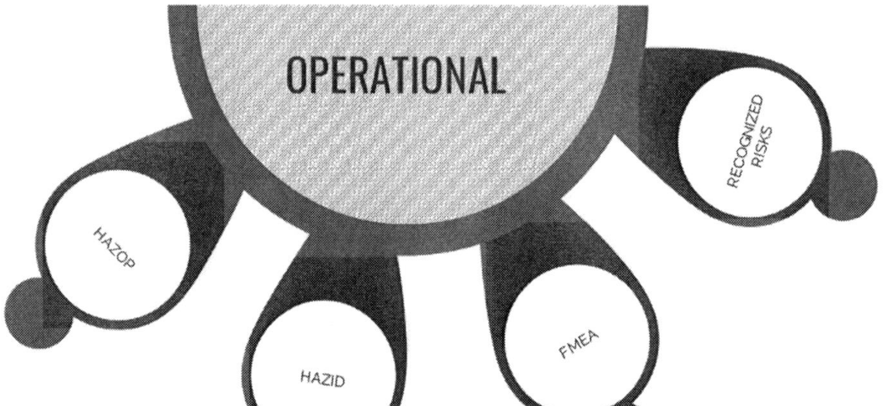

INTEGRATION GIVES MORE VALUABLE INFORMATION TO ENTERPRISE RISK MANAGERS ENSURING MUCH BETTER QUALITY OF TOP-DECISION MAKING

INTEGRATION GIVES THE OPERATIONAL RISK OWNERS GREATER VISIBILITY OF THEIR RISKS AND ASSURANCE THAT THEY WILL BE CONSIDERED

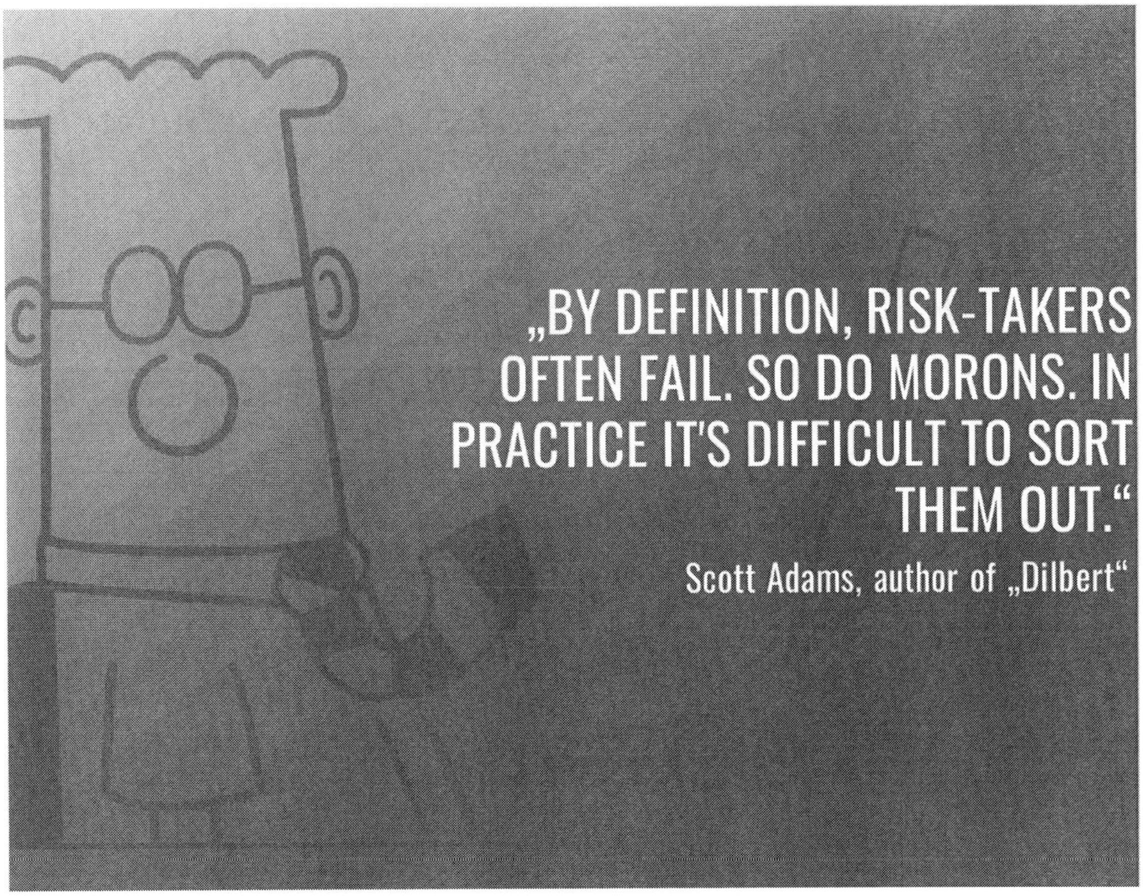

enablon

Thanks So Much For Listening

Integrating Operational And Enterprise Risk Management To Improve Asset Integrity And Reliability

Don't take our word for it. The project was recognized:

- **Enablon** Excellence in Technology Innovation SPF EMEA 2019, Winner
- **Verdantix** EHS Innovation Award 2019, Runner up

MOLGROUP

ERM

BIG DATA ANALYTIC AND EMERGENCY MANAGEMENT

BEGUM DOGAN: LABOR & SOCIAL SECURITY TRAINING EXPERT
CASGEM
www.casgem.gov.tr

OUTLINE

- Introduction
- Emergency management deficiencies in CSB Reports
- Results of CSB Report Analysis
- Concluding Remarks

Are Extremes Becoming More Frequents

Number of event
Relevant natural loss event
Worlwide 1980- 2018

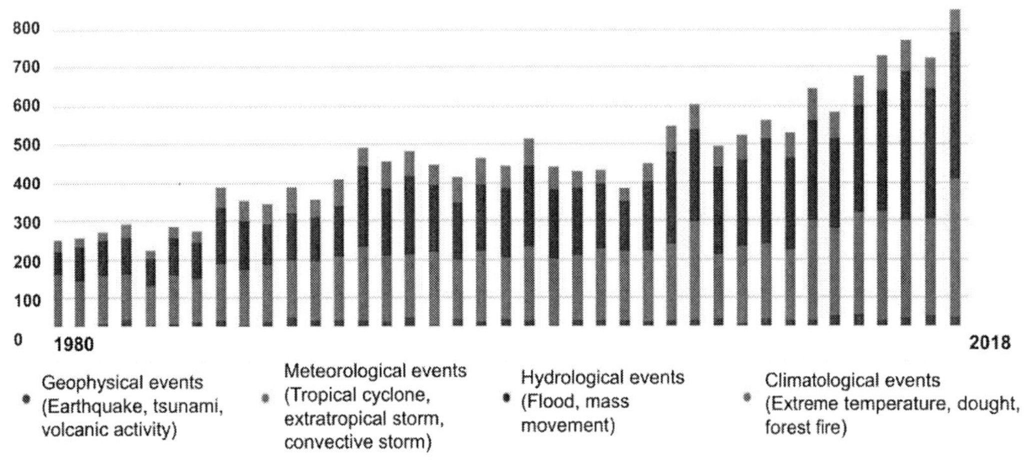

Are Extremes Becoming More Frequents

Relative trends of loss-relevant natural extreme events in different parts of the World

(1980 levels sea at 100 % data from Munich Re NatCatSERVICE)

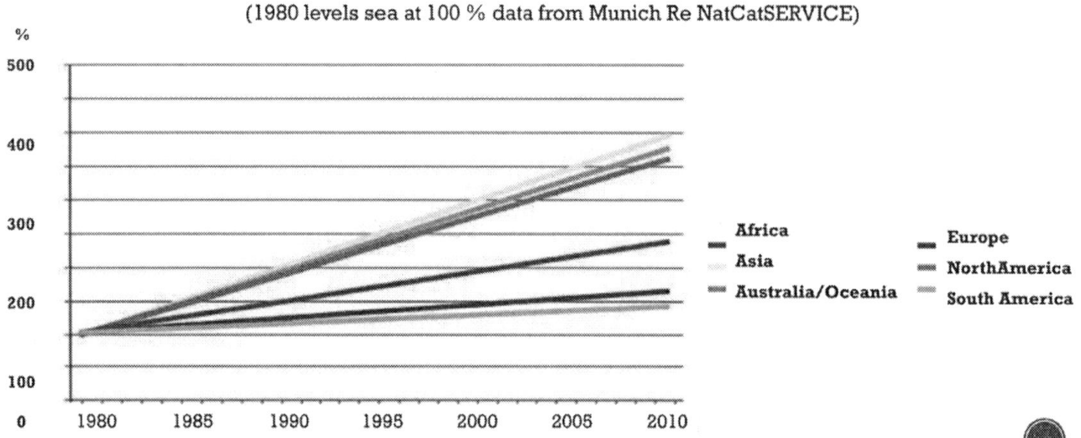

Reference:EASAC, Trends in extreme weather events in Europe: implications for national and European Union adaptation strategies

Benefits of Integration Big Data into Emergency Management

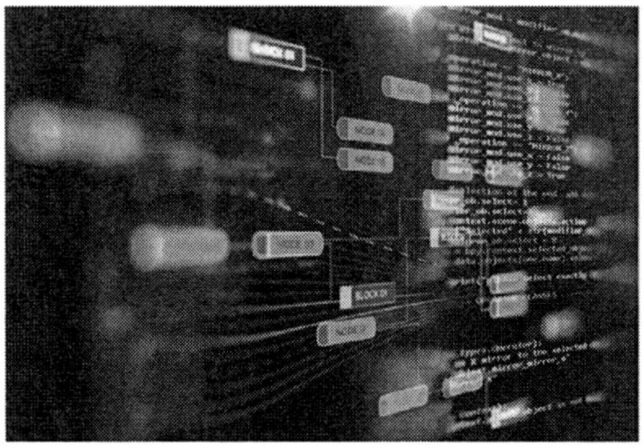

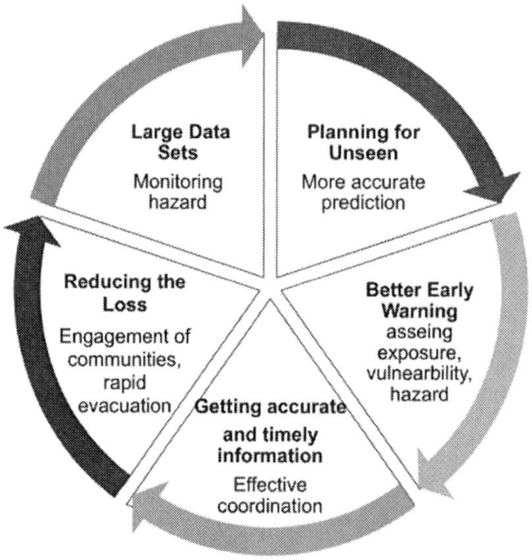

Challenges in Integration Big Data into Emergency Management

- Data privacy and data security
- Data storage, and resilience of big data system
- Filter failure problem
- Not a well know term for Policy makers, Emergency Department.
- Difficult to employee training keep up with technology
- Language and cultural differences

Emergency Management and Big Data in Articles

Data Sources	Method
Database	Machine Learning
Social media	Data mining and semantic technology
Real time data from sensor	Data mining on cloud computing
Oral ve textual data	Real time situation analysis, multimedia analysis, big data techniques

Emergency Management and Big Data in Articles

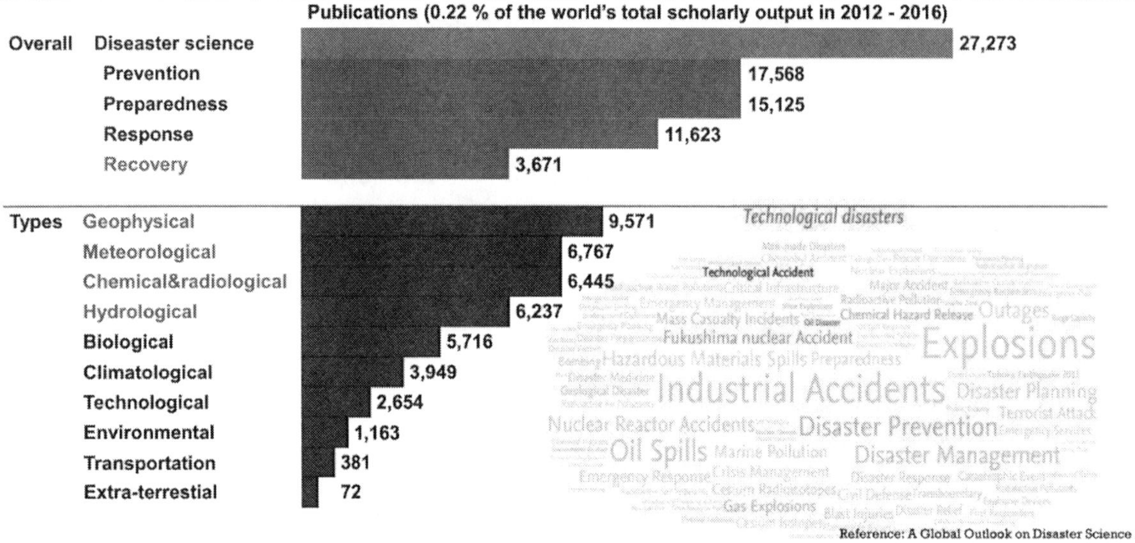

Reference: A Global Outlook on Disaster Science

Can we predict human emergency behavior and its movements by modeling its past movements during disaster?

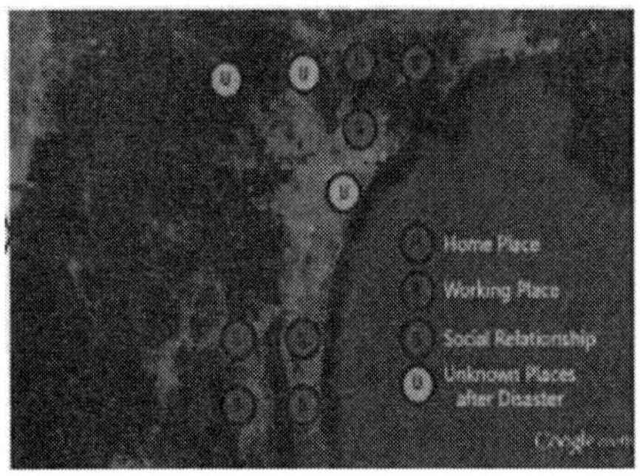

- GPS records of 1.6 million individuals Japan over 1 year.
- Several different dataset following the Great East Japan Earthquake and Fukushima nuclear accident.

- Hidden Markov model (HMM) to model dependency between disaster behaviors.
- Markov Decision Process (MDPs)

Reference: Song et al, Prediction of Human Emergency Behavior and their Mobility following Large-scale Disaster

The Four Phases of Emergency Management

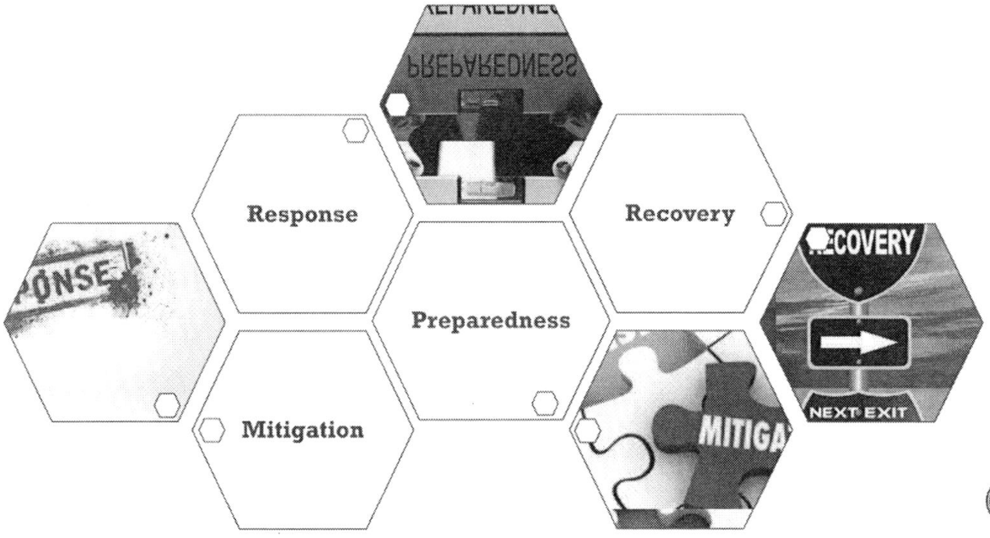

Organizations Fail to Learn Repeat Errors and Mistakes in Emergency Management «FEMA»

- Lack of viable plan
- Resource conflicts
- Lack of visible leadership
- Bad decisions make the situation worse
- Trying to obtain too much information, while neglecting the information flow
- Focusing on the insignificant
- Unknown Operations of staff

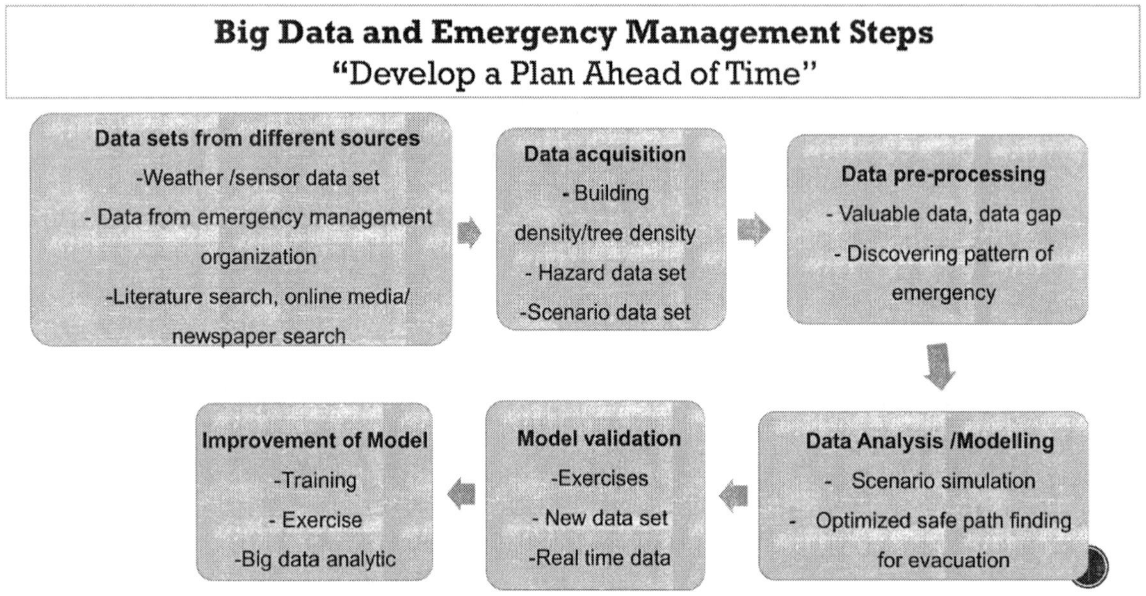
Big Data and Emergency Management Steps
"Develop a Plan Ahead of Time"
Data sets from different sources
-Weather /sensor data set
- Data from emergency management organization
-Literature search, online media/ newspaper search
Data acquisition
- Building density/tree density
- Hazard data set
-Scenario data set
Data pre-processing
- Valuable data, data gap
- Discovering pattern of emergency
Data Analysis /Modelling
- Scenario simulation
- Optimized safe path finding for evacuation
Model validation
-Exercises
- New data set
-Real time data
Improvement of Model
-Training
- Exercise
-Big data analytic

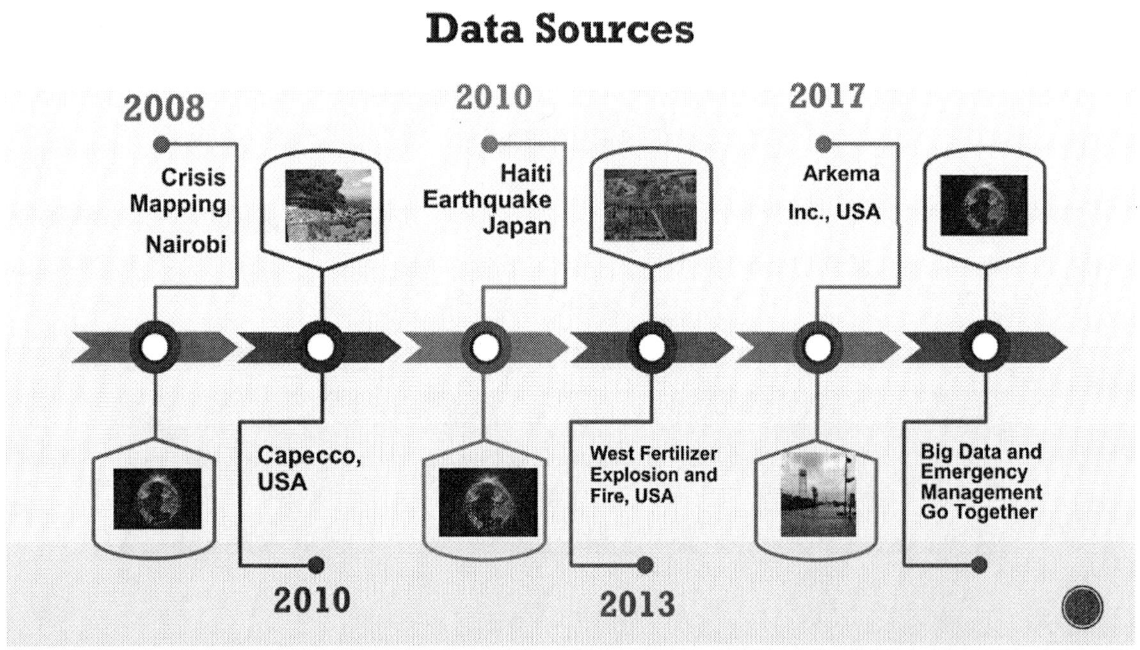
Data Sources
2008
Crisis Mapping Nairobi
2010
Capecco, USA
2010
Haiti Earthquake Japan
2013
West Fertilizer Explosion and Fire, USA
2017
Arkema Inc., USA
Big Data and Emergency Management Go Together

CSB Report Review (January 2009 – April 2019)

Keyword search, data set, emergency pattern

MITIGATION STAGE

- ✓ Not following hiyerarcy of control
- ✓ Lack of warning system OR Warning system failure
- ✓ Not using past data
- ✓ Lack of PSM
- ✓ Failure in risk assesment
- ✓ Lack of mitigation procedure/system
- ✓ Lack of incident management system
- ✓ Not provide sufficient PPE

PREPAREDNESS STAGE

- ✓ Lack of training, knowledge
- ✓ Lack of technical rescue certificitied emergency responder
- ✓ Lack of equipment, knowledge, certificated employee
- ✓ Failure in emergency plan
  - Bad scenario (not knowing risk)
  - Not exercising all scenario
  - Not implemented ERP across entire facility

CSB Report Review (January 2009 – April 2019)

FAILURE IN RESPONSE STAGE

FACILITIES

- ✓ Not following procedure, plan
- ✓ Coordination failure, conflict control
- ✓ Lack of established incident/evacuation management system
- ✓ Response procedure deficiencies
- ✓ Absence of safety path
- ✓ Unreliable/unreachable alarm system
- ✓ Not sharing information with responders

RESPONDER

- ✓ Arrive/scene late
- ✓ Lack of incident command system, coordination failure
- ✓ Bad decision
- ✓ Lack of equipment and PPE
- ✓ Lack of knowledge

Results

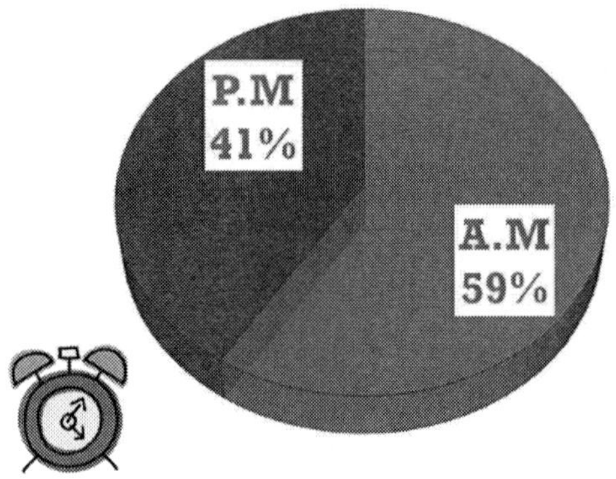

Failure in EM stage

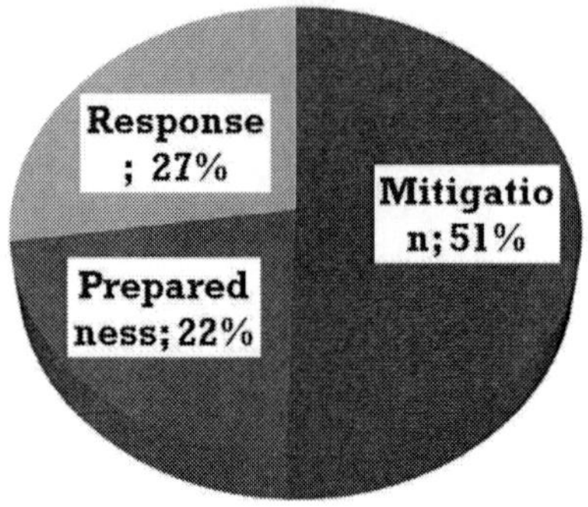

Results

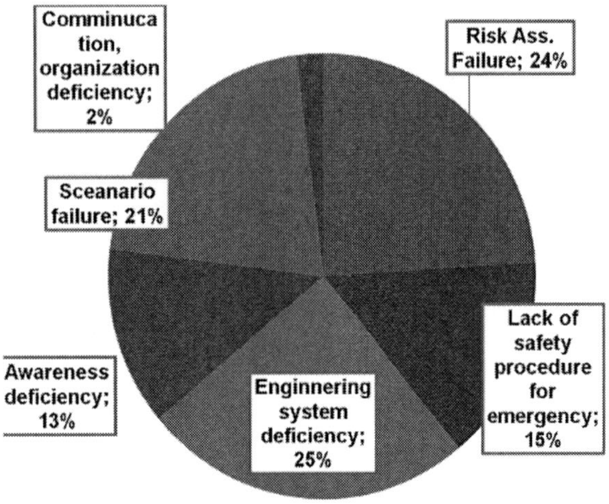

MITIGATON STAGE FAILURE

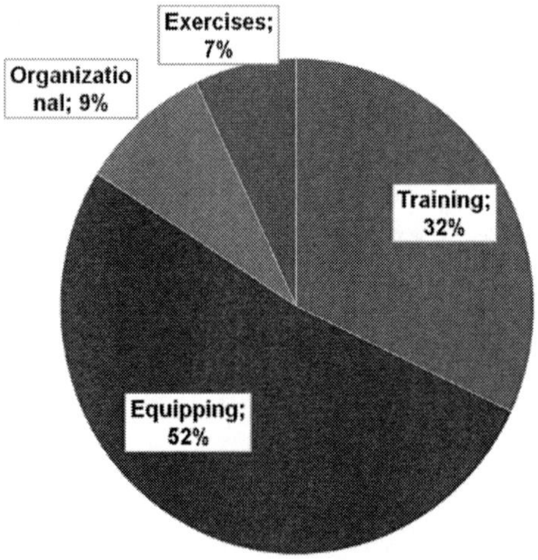

PREPAREDNESS STAGE FAILURE

Results & Conclusion Remarks

RESPONSE STAGE FAILURE

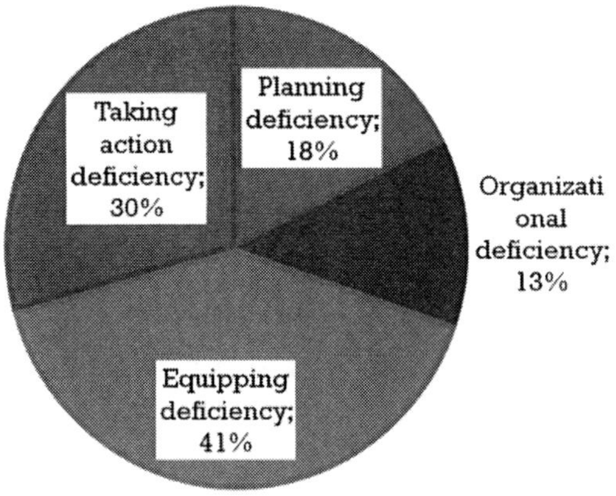

Deficiencies in Response Stage	%
Equipment /PPE deficiencies	33 %
Communication problem/ conflict control	19 %
Bad decision	19 %
Early warning system failure	10 %
Not following plan	7 %
Absence of evacuation plan	7 %
Absence of certified person or specialised process knowledge people	5 %

Thank you for listening....

Begum DOGAN
Labor & Social Security Training Expert
Ministry of Family, Labour and Social Service
begum.dogan@ailevecalisma.gov.tr
www.casgem.gov.tr

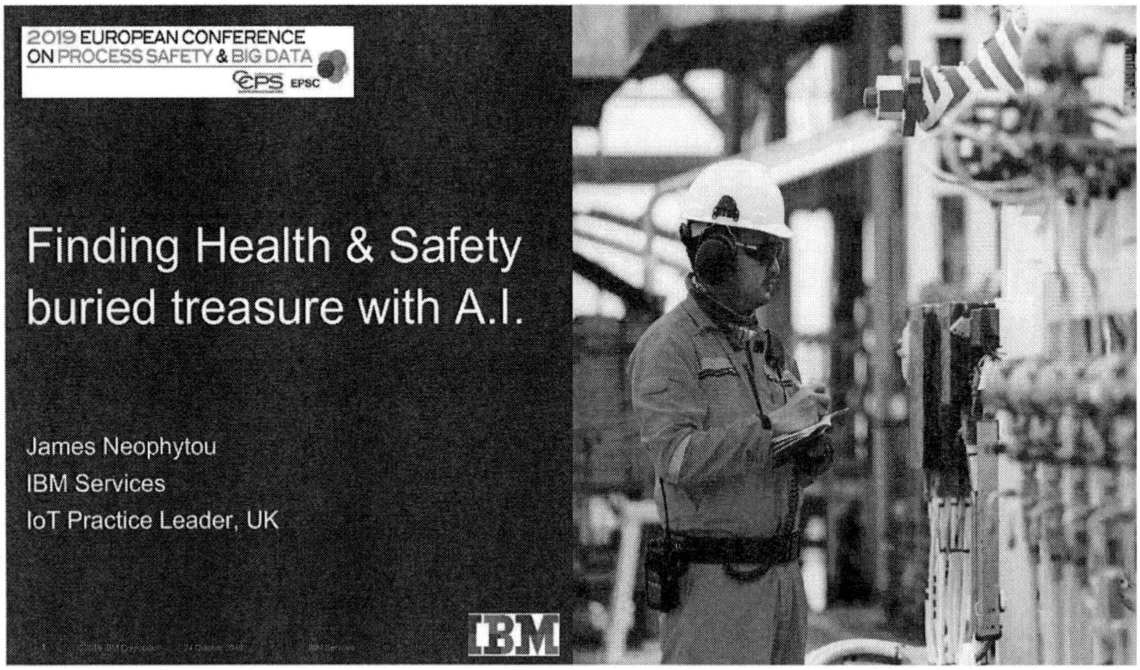
2019 EUROPEAN CONFERENCE
ON PROCESS SAFETY & BIG DATA
CCPS EPSC
Finding Health & Safety
buried treasure with A.I.
James Neophytou
IBM Services
IoT Practice Leader, UK
IBM

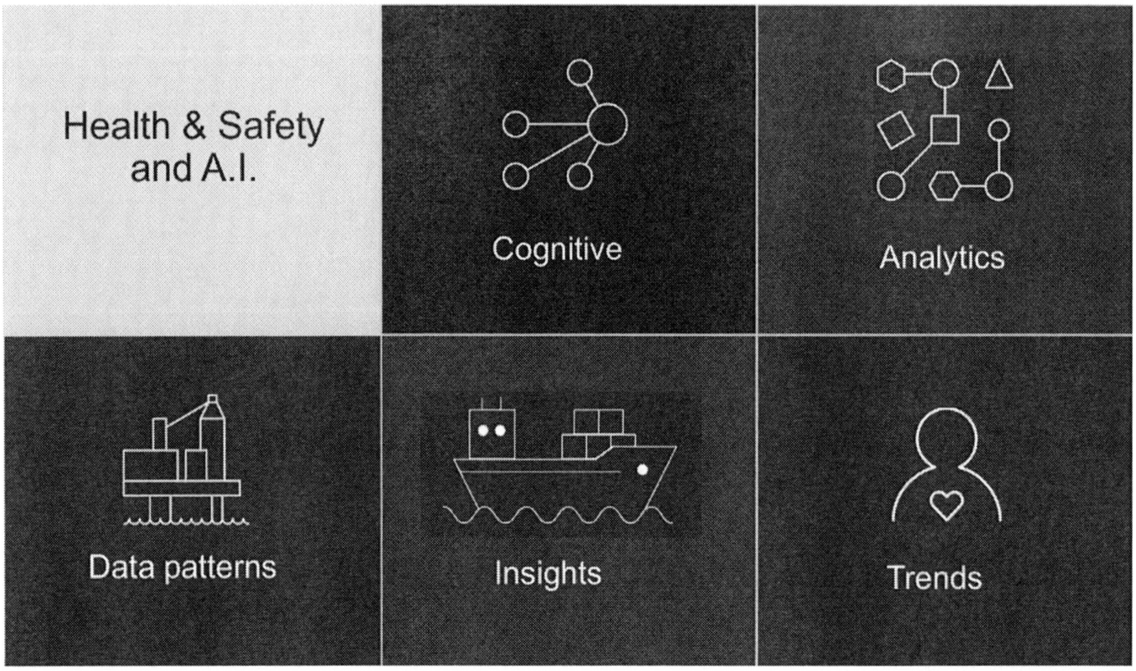
Health & Safety
and A.I.
Cognitive
Analytics
Data patterns
Insights
Trends

UK Water Company – **Lost Time Incidents** - Analytics

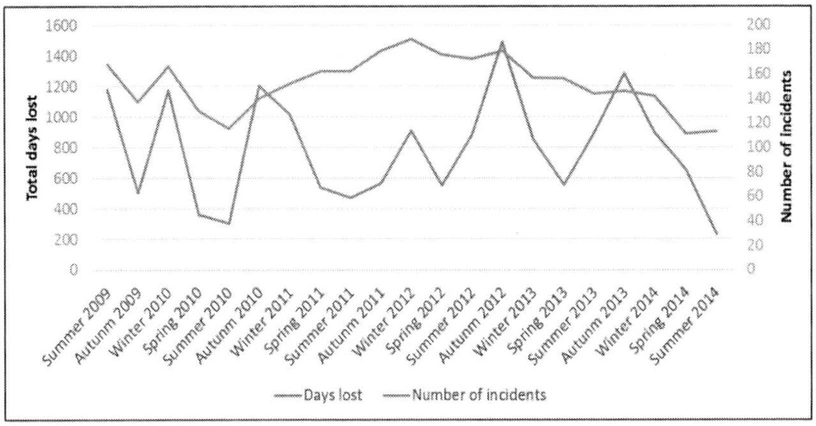

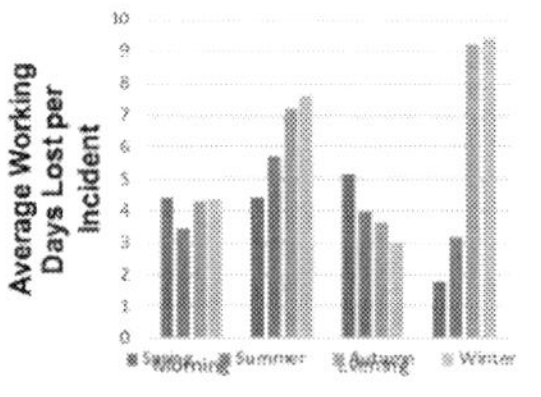

Total days lost varies with time | But number of incidents is steady

- Seasonal variations
- Time of day
- Changing behaviour
- Interventions and site visits

AI is a new approach

Cognitive systems forge a new partnership between humans and machines.

Humans excel at:
Common Sense
Morals
Imagination
Compassion
Abstraction
Dilemmas
Dreaming
Generalization

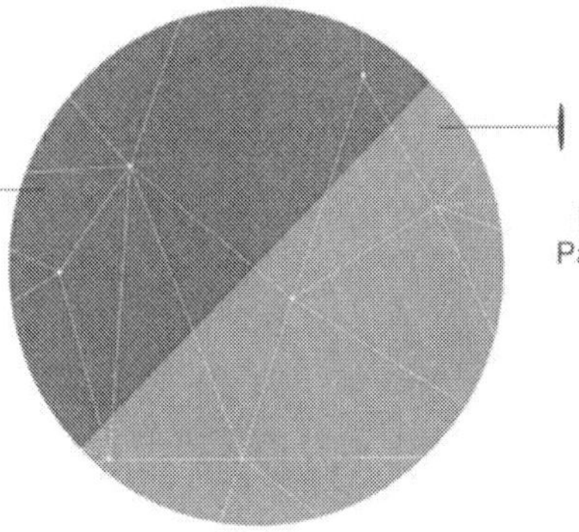

Cognitive Systems excel at:
Locating Knowledge
Pattern Identification
Natural Language
Machine Learning
Eliminate Bias
Endless Capacity

AI "Senses"

Seeing, hearing, speaking, finding, feeling, interpreting

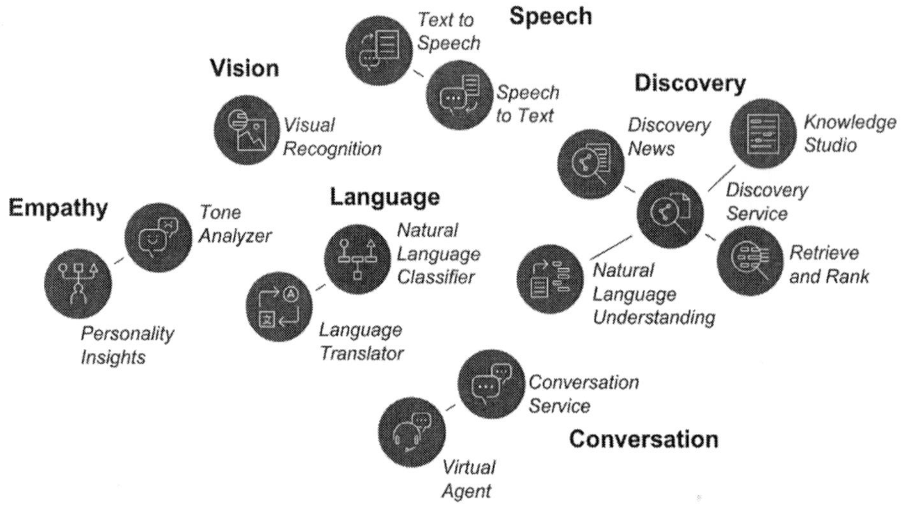

AI Family

Artificial intelligence
Simulation of human intelligence processes

Recommendation engines
Analyze data and suggest something as per user's interest

Robotics
Conception, design, manufacture, and operation of robots

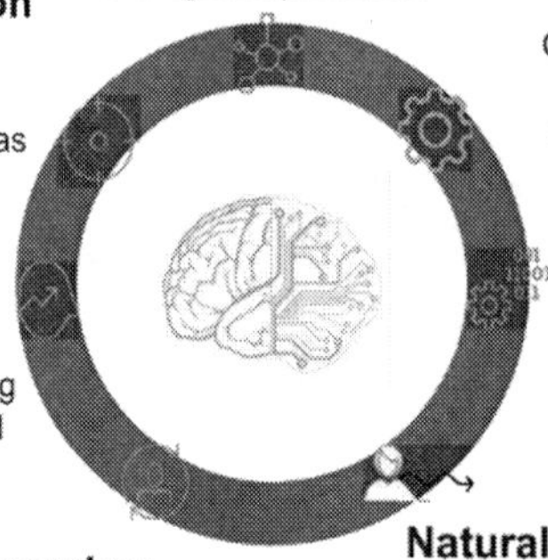

Machine learning systems
Ability to learn and improve without explicit instructions

Predictive analytics
Predicting outcomes using statistical algorithms and machine learning

Deep learning
Machine learning with artificial neural network algorithms

Natural language processing
Ability to understand human speech as it is spoken

The value for HSE

Pain Points

- Declining operational excellence performance
- **Lost knowledge** from workforce attrition
- **Low sharing** of incident information across sites
- Lessons learned from past events **forgotten**
- **Reinventing the wheel** when taking actions to mitigate similar risks
- **Intensive manual effort** to gather and research incident information, every time a hazard study is performed
- **Locating similar events** within historical archives
- **Untapped knowledge** buried in text or documents

AI Benefits

- Explore documents and unstructured text for **incident insights, using natural language**
- Immediately locate **similar incident reports** based on common themes and concepts
- **Automatically find** key items such as hazard conditions, activities, incident causes and controls
- Uncover **hidden trends** in behaviors and undetected hazards
- Reduce reliance on collective corporate memory
- Scale and augment HSE expertise
- Offer a new toolkit for Operational improvement

Which questions are too hard today?

- Find the **10 most recent incidents similar to** incident X
- Find **all incidents involving** carbon steel piping
- What are the **most common activities** involved with serious accidents?
- What hazard conditions are encountered **most frequently at site X**?
- What **activities in the company are the most risky**? Is this corroborated in public data?
- Where is training indicated as a recommended co**ntrol most often**?
- I will be performing tank inspections next week at site Y. What are the **recommended risk controls** on the most recent confined space incidents there?

Effort involved in studies and investigations

- Find an incident report
- Read it through
- Determine key information
- Determine relevance to current study
- Find something similar elsewhere
- Rinse and repeat.
- Do it again next time.

Where do I find Buried Treasure?

The following shows real-life examples from a working AI model.

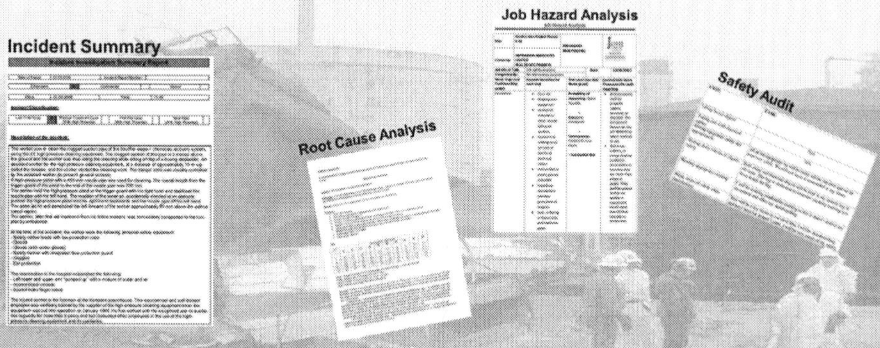

The model is based on treasure found at these sites:-

Workplace Health & Safety Queensland, AUS
Pacific Northwest National Laboratory H2 Tools, USA
International Association of Oil & Gas Producers, UK
Australian Petroleum Production & Exploration Association, AUS
National Offshore Petroleum Safety and Environmental Management Authority

What Types of **Equipment** were involved?

NOPSA
NATIONAL OFFSHORE PETROLEUM SAFETY AUTHORITY

SAFETY ALERT 25

Tensioner Failure

What happened?

In Australian waters, two separate incidents have occurred recently on semi-submersible MODUs. Each incident involved a guide line or pod line tensioner operated Rod End Down (RED). In each case, the wire rope parted, releasing the load on the tensioner and allowing the tensioner piston to stroke downwards in an unrestrained manner until it came into contact with the end cover plate of the cylinder. The impact of the piston on the end of the cylinder was severe enough to cause damage. In the first case the mounting bolts fractured allowing the complete tensioner unit to fall to the deck. In the second case, the bottom cover plate securing bolts fractured, allowing the piston assembly to come out of the cylinder and land on the grating below.

Both of these incidents had the potential to cause fatal injuries if persons had been in the vicinity.

In Australian waters, two separate incidents have occurred recently on semi-submersible MODUs.

Each incident involved a guide line or pod line tensioner operated Rod End Down (RED).

In each case, the wire rope parted, releasing the load on the tensioner and allowing the tensioner piston to stroke downwards in an unrestrained manner until it came into contact with the end cover plate of the cylinder.

The impact of the piston on the end of the cylinder was severe enough to cause damage.

In the first case the mounting bolts fractured allowing the complete tensioner unit to fall to the deck.

In the second case, the bottom cover plate securing bolts fractured, allowing the piston assembly to come out of the cylinder and land on the grating below.

Both of these incidents had the potential to cause fatal injuries if persons had been in the vicinity.

https://www.nopsema.gov.au/assets/Safety-alerts/Alert25.pdf

What **Activities** were being performed?

IOGP high potential event report

Incident description:

IP was hit by scaffolding pipe tip during Lifting the Scaffolding from Fire Water Pump Caisson During Fire Water Pump caisson repair work, a high-pressure jet hose was used to clean-up the cement inside the caisson after completion of the grouting job. The hose was tied at the end of 5 pieces scaffolding pipe so it could be lowered down into the caisson. After water jetting was completed, an attempt was made to pull-out the scaffolding pipe from the caisson by crane. The first piece of the scaffolding pipe was loose and left the remaining pipe hung inside the caisson. It was decided to pull-out the remaining scaffolding by connecting the hook to the high-pressure hose. During subsequent recovery of the pipe, when the IP was attempting to look down inside the caisson the tip of the scaffolding came up and hit him on the nose.

http://safetyzone.iogp.org/HighPotentialEvents/detail.asp?inc_id=279

Material

IP was hit by scaffolding pipe tip during Lifting the Scaffolding from Fire Water Pump Caisson During Fire Water Pump caisson repair work, a high-pressure jet hose was used to clean-up the cement inside the caisson after completion of the grouting job.

The hose was tied at the end of 5 pieces scaffolding pipe so it could be lowered down into the caisson.

After water jetting was completed, an attempt was made to pull-out the scaffolding pipe from the caisson by crane.

The first piece of the scaffolding pipe was loose and left the remaining pipe hung inside the caisson.

It was decided to pull-out the remaining scaffolding by connecting the hook to the high-pressure hose

.

During subsequent recovery of the pipe, when the IP was attempting to look down inside the caisson the tip of the scaffolding came up and hit him on the nose.

Equip Type

Body part

IBM

What **Hazards** were encountered?

High Potential Incident Alert - 129

June 2013

Hydrocarbon release – Pressure Safety Valve chatters after failing to reseat

INCIDENT SUMMARY

A gas release was observed during the restart of a platform following a planned shutdown.

The platform was immediately shut down and de-pressured.

No personnel were injured during this event.

appea

CRITICAL FACTORS

1. Pressure spike in system lifts the Pressure Safety Valve (PSV).
2. PSV failed to reseat because the reseat pressure was lower than the operating pressure.
3. Extended duration of PSV chatter, led to eventual failure of the system weak point (pipe support) and the hydrocarbon release.

KEY LESSONS

- Consider all conditions in PSV selection during detailed design – conventional PSV blowdown is ~10%. Under 2-Phase flow conditions this can be as high as 20% and needs to be considered in PSV selection during detailed design.

https://www.appea.com.au/wp-content/uploads/2013/09/129-Hydrocarbon-release-Pressure-Safety-Valve-chatters-after-failing-to-reseat.pdf

A gas release was observed during the restart of a platform following a planned shutdown.

The platform was immediately shut down and de- pressured.

No personnel were injured during this event.

FINDINGS: 1. Pressure spike in system lifts the Pressure Safety Valve (PSV).

2.

PSV failed to reseat because the reseat pressure was lower than the operating pressure.

3. Extended duration of PSV chatter, led to eventual failure of the system weak point (pipe support) and the hydrocarbon release.

ACTIONS: > Consider all conditions in PSV selection during detailed design conventional PSV blowdown is ~10%.

Under 2-Phase flow conditions this can be as high as 20% and needs to be considered in PSV selection during detailed design.

Equip Type Activity

IBM

What **Causes** and **Controls** were identified?

- Management Systems
- Work Direction
- Procedures
- Human Factors
- Human Engineering
- Communications
- Equipment
- Quality Control
- Training

- Engineering/Design
- Maintenance/Inspection
- Mitigating Devices
- Warning Devices
- Training/Procedures
- Personal Protection

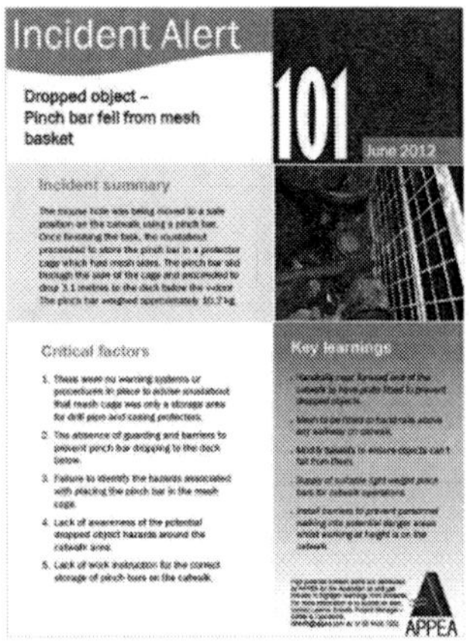
Incident Alert

Dropped object –
Pinch bar fell from mesh basket

101

June 2012

Incident summary

The mouse hole was being moved to a safe position on the catwalk using a pinch bar. Once finishing the task, the roustabout proceeded to store the pinch bar in a protector cage which has mesh sides. The pinch bar slid through the side of the cage and proceeded to drop 3.1 metres to the deck below the v-door. The pinch bar weighed approximately 10.7 kg.

Critical factors

1. There were no warning systems or procedures in place to advise roustabout that mesh cage was only a storage area for drill pipe and casing protectors.
2. The absence of guarding and barriers to prevent pinch bar dropping to the deck below.
3. Failure to identify the hazards associated with placing the pinch bar in the mesh cage.
4. Lack of awareness of the potential dropped object hazards around the catwalk area.
5. Lack of work instruction for the correct storage of pinch bars on the catwalk.

Key learnings

APPEA

https://www.appea.com.au/wp-content/uploads/2013/04/101-Dropped-object-Pinch-bar-fell-from-mesh-basket.pdf

FINDINGS: 1. There were no warning systems or procedures in place to advise roustabout that mesh cage was only a storage area for drill pipe and casing protectors.

2. The absence of guarding and barriers to prevent pinch bar dropping to the deck below.

3. Failure to identify the hazards associated with placing the pinch bar in the mesh cage.

4. Lack of awareness of the potential dropped object hazards around the catwalk area.

5. Lack of work instruction for the correct storage of pinch bars on the catwalk.

ACTIONS: > Handrails near forward end of the catwalk to have plate fitted to prevent dropped objects.

> Mesh to be fitted to hand rails above any walkway on catwalk.

> Modify baskets to ensure objects can't fall from them.

> Supply of suitable light weight pinch bars for catwalk operations.

> Install barriers to prevent personnel walking into potential danger areas whilst working at height is on the catwalk.

Equip Type | Activity | Hazard

IBM

Let's put it all together – Illuminating the Insights

IOGP high potential event report

Incident description:

A machine toppled to the left and came to rest over the edge of the secondary pipe rack excavation, missing the pedestal. The approximate height from where they were working to where they came to rest was 8 metres. Thanks the appropriate fall arrest system and the integrity of the basket guarding no one was seriously injured.

What Went Wrong:

• Unsafe work practices: Manual lifting of handrail from man riding basket, operating man- lift at the limit - positio beyond horizontal outreach as per data sheet. • Violating of HSSE procedures: Use of unauthorised lifting equipme (pulley and rope) • Lack of work control: Inadequate PTW for man-riding and working at height. (Generic PTW for construction at pipe rack covers lifting with 70t and 500t crane but not man-riding and working at height. • Hazar and risks - failure to recognize hazards: Although a task risk assessment was developed, some hazards and risks we neglected. Although the hazards of unauthorised lifting equipment were identified and immediately acted upon at incident site, the unsafe work practices during man-riding and working at height remain unrecognized. • Hazards : risks - failure to communicate and manage hazards and risks: There is no evidence that hazards and mitigation measures relevant to man-riding and working at height have been adequately communicated at the day of the inc (tool box talk) • Possible malfunctioning of man-lift limit switches (actual vertical extension of lam possible, versus maximum extension as per man- lift data sheet).

Corrective Actions:

* Ensure that unauthorised lifting equipment must not be used through increased HSSE supervision and 0-toleran policy. * Ensure that tasks involving man riding and working at height are controlled by separate permits and task assessments considering amongst others: adequacy and positioning of equipment (permits to be issued and valid on daily basis). * Contractor HSSE supervisors to ensure that man riding and working at height hazards and the us authorised lifting equipment are discussed at daily TBTs. * Ensure that "stop work" instructions for unsafe activities followed by subcontractors without further discussions. * Refresher training on man riding and working at height t organised for man-lift operators and supervisors. * Conduct 3rd party inspection at MEWPs in order to verify the function of the safety features (in particular limit switches). * Ensure that permanent supervision and an area inspe by the HSE coordinator is carried out for work involving man-riding and working at height. * Improve the coordina between mechanical and civil work to ensure proper access of cranes and man-riding equipment to the work site. Senior managers need to be present at the project construction sites for walk arounds discussing high risk activitie golden rules. In particular, "working at height", lifting and "stop work". * Contractor supervision: construction team to focus on the supervision of activities involving working at height, man-riding and lifting by: - Stopping unsafe w Verifying task risk assessment, permits, and certificates. - Verifying use of proper lifting equipment - Accessibility c work site with lifting equipment. - Usage of adequate PPEs. * Issue a safety alert

http://safetyzone.iogp.org/HighPotentialEvents/detail.asp?inc_id=2803

A machine toppled to the left and came to rest over the edge of the secondary pipe rack excavation, missing the pedestal.

The approximate height from where they were working to where they came to rest was 8 metres.

Thanks to the appropriate fall arrest system and the integrity of the basket guarding no one was seriously injured.

FINDINGS: • Unsafe work practices: Manual lifting of handrail from man riding basket, operating man-lift at the limit - position beyond horizontal outreach as per data sheet.

• Violating of HSSE procedures: Use of unauthorised lifting equipment (pulley and rope) • Lack of work control: Inadequate PTW for man-riding and working at height.

(Generic PTW for steel construction at pipe rack covers lifting with 70t and 500t crane but not man-riding and working at height.

• Hazards and risks - failure to recognize hazards: Although a task risk assessment was developed, some hazards and risks were neglected.

Although the hazards of unauthorised lifting equipment were identified and immediately acted upon at the incident site, the unsafe work practices during man-riding and working at height remain unrecognized.

• Hazards and risks - failure to communicate and manage hazards and risks: There is no evidence that hazards and mitigation measures relevant to man-riding and working at height have been adequately communicated at the day of the incident (tool box talk) • Possible malfunctioning of man-lift limit switches (actual vertical extension of lam possible, versus 14 m maximum extension as per man-lift data sheet).

ACTIONS: * Ensure that unauthorised lifting equipment must not be used through increased HSSE supervision and 0-tolerance policy.

* Ensure that tasks involving man riding and working at height are controlled by separate permits and task risk assessments considering amongst others: adequacy and positioning of equipment (permits to be issued and validated on daily basis).

* Contractor HSSE supervisors to ensure that man riding and working at height hazards and the use of authorised lifting equipment are discussed at daily TBTs.

* Ensure that "stop work" instructions for unsafe activities are followed by subcontractors without further discussions.

* Refresher training on man riding and working at height to be organised for man-lift operators and supervisors.

* Conduct 3rd party inspection at MEWPs in order to verify the function of the safety features (in particular limit switches).

* Ensure that permanent supervision and an area inspection by the HSE coordinator is carried out for work involving man-riding and working at height.

* Improve the coordination between mechanical and civil work to ensure proper access of cranes and man-riding equipment to the work site.

* Senior managers need to be present at the project construction sites for walk arounds discussing high risk activities and golden rules.

In particular, "working at height", lifting and "stop work".

* Contractor supervision: construction team HSE to focus on the supervision of activities involving working at height, man-riding and lifting by: - Stopping unsafe work - Verifying task risk assessment, permits, and certificates.

- Verifying use of proper lifting equipment - Accessibility of work site with lifting equipment.

- Usage of adequate PPEs.

* Issue a safety alert

IBM

Finding New Insights

The relief valve on the line between the turbocharger and the methanation reactor was causing it to open without the operator noticing.

New Human Factor cause

https://h2tools.org/lessons/hydrogen-leak-ignites-and-explodes-ammonia-production-plant

Inadequate management of community expectations in a hostile working environment.

New Management System cause

http://safetyzone.iogp.org/HighPotentialEvents/detail.asp?inc_id=2658

FINDINGS: DVT is a blood clot in one of the deep veins usually in the calf or thigh.

In this part of the body, blood moves quite slowly and when a clot occurs, it can get stuck in th supply.

Symptoms aren"t always immediately obvious and a blood clot is not necessarily fatal.

However, these blood clots can become dislodged and travel to the lungs or other areas, caus damage, or death.

When you sit for a long time without contracting the muscles in your legs, blood can pool in th

New "blood clot" and "blood can pool" hazards

https://www.nopsema.gov.au/assets/Safety-alerts/Alert27.pdf

Experience transfer represents the foundation

New Training/Procedures control

http://safetyzone.iogp.org/HighPotentialEvents/detail.asp?inc_id=2644

Gas detection not used near pump unit or near tanks.

New Work Direction cause

http://safetyzone.iogp.org/HighPotentialEvents/detail.asp?inc_id=2652

Finding Similar Incidents

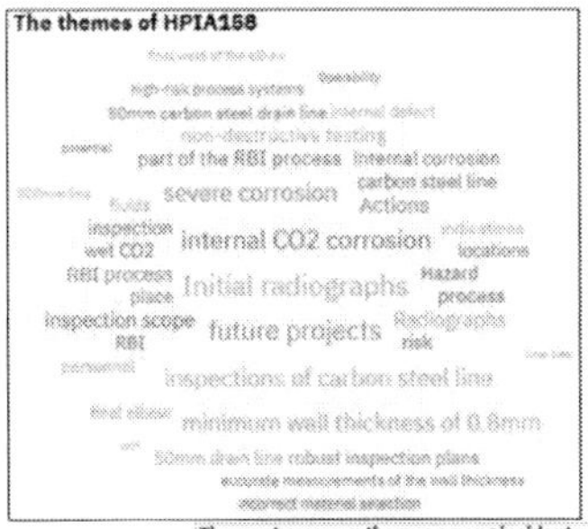

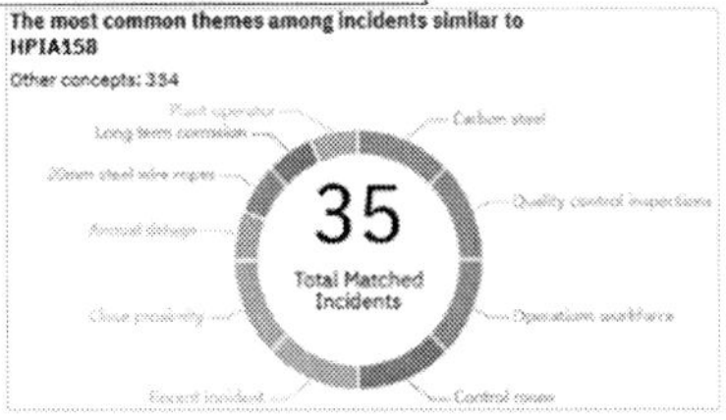

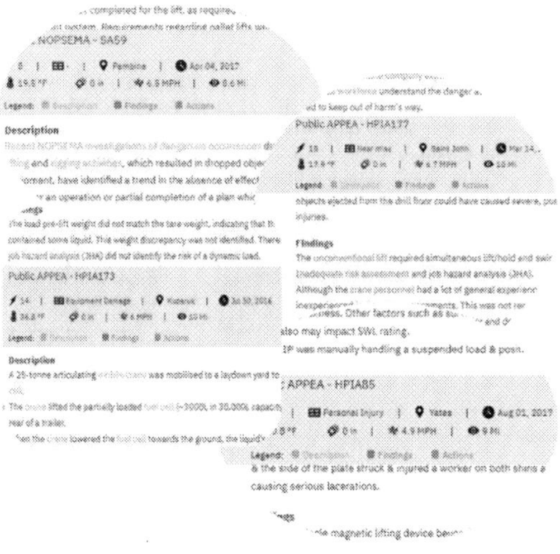

Teaching AI the language of Safety: Bow-Tie Risk model

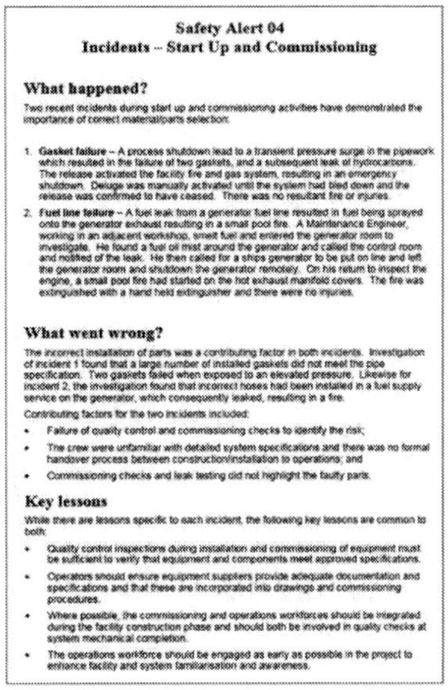

Safety Alert 04
Incidents – Start Up and Commissioning

What happened?

Two recent incidents during start up and commissioning activities have demonstrated the importance of correct material/parts selection:

1. **Gasket failure** – A process shutdown lead to a transient pressure surge in the pipework which resulted in the failure of two gaskets, and a subsequent leak of hydrocarbons. The release activated the facility fire and gas system, resulting in an emergency shutdown. Deluge was manually activated until the system had bled down and the release was confirmed to have ceased. There was no resultant fire or injuries.
2. **Fuel line failure** – A fuel leak from a generator fuel line resulted in fuel being sprayed onto the generator exhaust resulting in a small pool fire. A Maintenance Engineer, working in an adjacent workshop, smelt fuel and entered the generator room to investigate. He found a fuel oil mist around the generator and called the control room and notified of the leak. He then called for a ships generator to be put on line and left the generator room and shutdown the generator remotely. On his return to inspect the engine, a small pool fire had started on the hot exhaust manifold covers. The fire was extinguished with a hand held extinguisher and there were no injuries.

What went wrong?

The incorrect installation of parts was a contributing factor in both incidents. Investigation of incident 1 found that a large number of installed gaskets did not meet the pipe specification. Two gaskets failed when exposed to an elevated pressure. Likewise for incident 2, the investigation found that incorrect hoses had been installed in a fuel supply service on the generator, which consequently leaked, resulting in a fire.

Contributing factors for the two incidents included:

- Failure of quality control and commissioning checks to identify the risk;
- The crew were unfamiliar with detailed system specifications and there was no formal handover process between construction/installation to operations; and
- Commissioning checks and leak testing did not highlight the faulty parts.

Key lessons

While there are lessons specific to each incident, the following key lessons are common to both:

- Quality control inspections during installation and commissioning of equipment must be sufficient to verify that equipment and components meet approved specifications.
- Operators should ensure equipment suppliers provide adequate documentation and specifications and that these are incorporated into drawings and commissioning procedures.
- Where possible, the commissioning and operations workforces should be integrated during the facility construction phase and should both be involved in quality checks at system mechanical completion.
- The operations workforce should be engaged as early as possible in the project to enhance facility and system familiarisation and awareness.

▸AI ▶

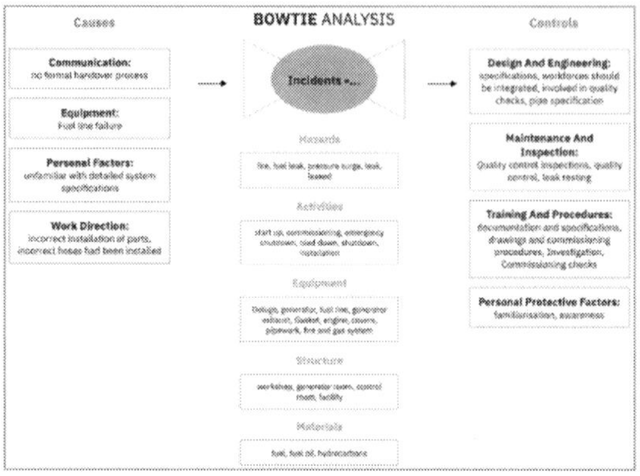

AI trends

NASA: Automated discovery of previously unknown precursors to aviation safety incidents

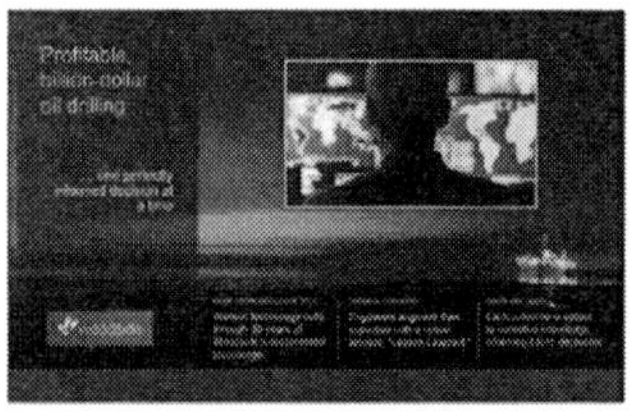

Woodside Energy: Use machine learning for 30 years of engineering data – Watson for Safety

https://youtu.be/GFZ2laTVkY8

Rail Safety & Standards Board: Making Big Data Risk Analysis work for the GB Railways

  24 October 2019 IBM Services IBM

Emerging Technologies – Source: Gartner Hype Cycle

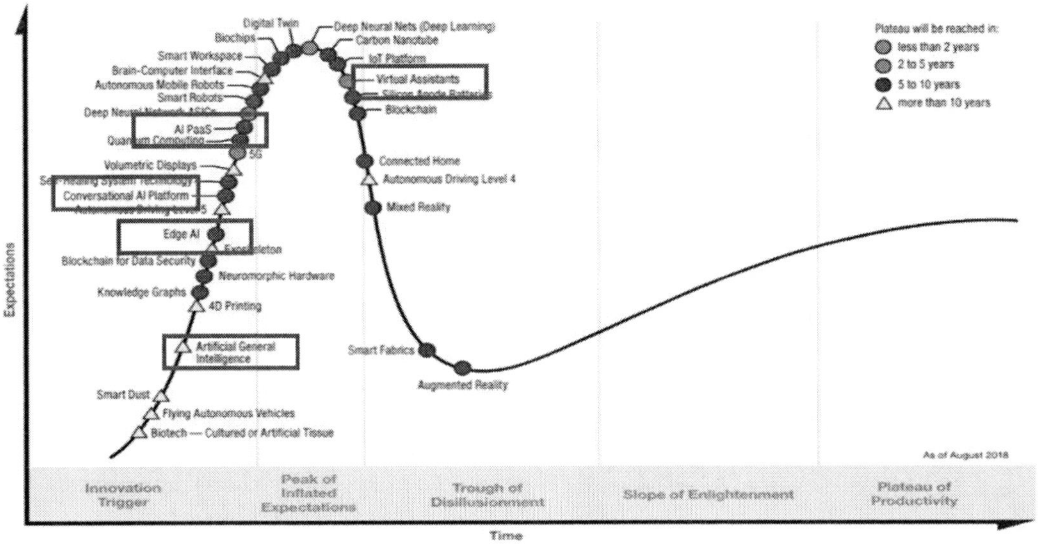

Launching an AI Programme

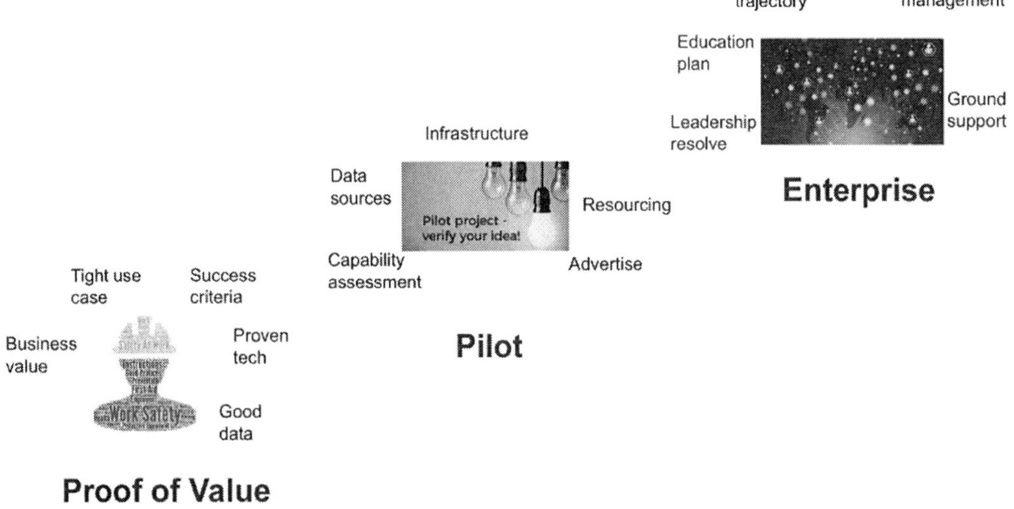

IBM

A Path to Prediction

Lagging Indicators: re-active; measures barrier defects; easy to measure

Leading Indicators: pro-active; measures the strength of barriers; difficult to measure

Latent Indicators: predictive; hidden measurements, undetected patterns; hard to discern due to unstructured nature; measures application and effectiveness of barriers over time

Where AI is especially suited for analysis and identification

- Helps to assess potential future risks based on past activities and conditions
- Looks at other unstructured data sources
- Helps to identify trends and new risks

Applying Analytics to the Process Safety Life-Cycle: Reveal and Repair Process Safety Vulnerabilities

Nicholas Sands, Senior Manufacturing Technology Fellow, DuPont

A.M. (Tony) Downes, Global Process Safety Advisor, Honeywell

Nicholas Sands

Senior Manufacturing Technology Fellow at DuPont

~30 years of Automation and Process Safety experience

Roles

- Global Alarm Management Leader for DuPont
- Currently working in Luxembourg
- Currently living in Dallas, Texas
- Certified Automation Professional
- International Society of Automation (ISA) Fellow
- Process Automation Hall of Fame
- Lead for many Automation Best Practices in DuPont

ISA/IEC STANDARDS

- Co-Chair of ISA18 on Alarm Management
- Co-Director of ISA84 on Safety Instrumented Systems
- Co-Director of ISA101 on Human Machine Interface
- Secretary of IEC62682 committee on AM, lead editor
- Member of other ISA standards: ISA105, ISA108…
- Editor of Guide to Automation Body of Knowledge 3rd edition

A little about the presenter

A.M. (**Tony**) **Downes**
Global Process Safety Advisor

1978-1988 **DuPont** Canada. Project Eng, Process Eng, Product Development, Maintenance Eng, Project Manager
1988-1992 **Bayer** Canada, Supervisor LPE
1992-2001 **Westlake** Group, Principle PS Eng.
2001-2010 **FMC**. Global Safety & Sec. Mgr
2010-date **Honeywell PMT** Global PS Advisor

- Led over 100 PHAs
- Did first LOPA in 1999
- Led over 100 Incident Investigations
- Launched 4 Risk Reduction programs

- CCPS Tech Steering and Planning Cmtes
- CCPS Certified Process Safety Engineer

"Everything I know about Process Safety, I learned in an investigation"

Honeywell
THE POWER OF CONNECTED

Agenda

- Safety Moment
  - Bypassing is Bad
- The Requirements
- The Architecture
- The Data
  - Case Study 1: Visibility→Improvement
  - Case Study 2: Bypassing SIL3
- The Benefits
- The Future
- Summary

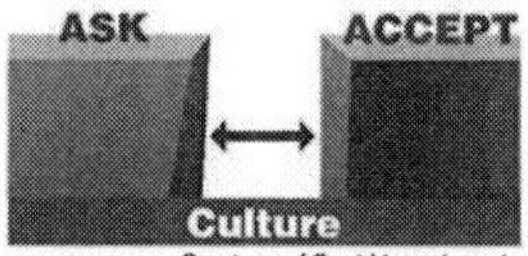

Courtesy of Fred Henselwood

10/16/2019 4

Aftermath of the BLEVE

What Happened

- In 1999, a partial power outage (in this case an electrical failure without loss of steam pressure) led to a catastrophic overpressure in several pressure vessels in an alumina slurry digestion/flash system.
  - Boiling Liquid, Expanding Vapor Explosion (BLEVE).
  - shockwave and the hot caustic liquid injured 29 people – several permanently
  - Damage was in the tens of millions of dollars.
  - Fortunately no one was killed.

Lessons

- Safety-related protections shouldn't be bypassed. Temporary MOCs may be used to manage bypasses for a short time while something is repaired, so long as you take other measures to ensure you aren't taking on more risk than the company intended to.
- If you regularly have to operate with critical safeguards bypassed or impaired, you should report this. Leadership may not know.

Read more in PS Beacon

Bypassing Safety Critical Safeguards should be Well-Controlled & Rare

Recognition

- Adrienne Ashley – Manufacturing Technology Principle Engineer, DuPont
  - Data and learnings
- Tony
  - Case 2

DUPONT

The Requirement

SIF demand tracking became requirement in Good Engineering Practices (International norms)

- IEC 61511 Ed1 2003
- ANSI/ISA S-84 2004
- IEC 61511 Ed 2 2015

IEC 61511-1 Edition 1

- 16.2.2 Operation and maintenance procedures shall be developed in accordance with the relevant safety planning and shall provide the following:
- …
- the information which needs to be maintained on system failure and demand rates on the SIS;
- 16.2.6 Discrepancies between expected behaviour and actual behaviour of the SIS shall be analysed and, where necessary, modifications made such that the required safety is maintained. This shall include monitoring the following:
- …
- the cause of the demands;
- the cause of false trips.

DUPONT

The Architecture

Demand tracking application:

- **enterprise architecture**
- **internal cloud environment**
- **secure data transport**
- **Since ~2010**

At the peak, ~45 plants were connected.

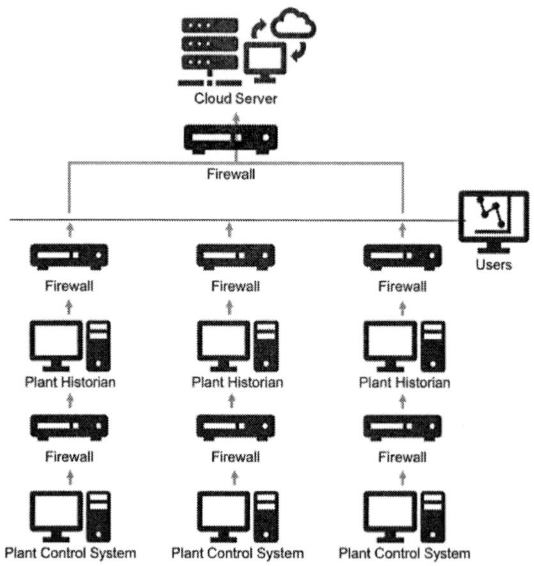

DUPONT

The Data: Case 1 Fewer Demands = Lower Risk

Data in a central server supported unit comparison and standardized reporting.

Data transparency enabled units to see and fix issues

Data transparency also raises profile & *Demands attention.*

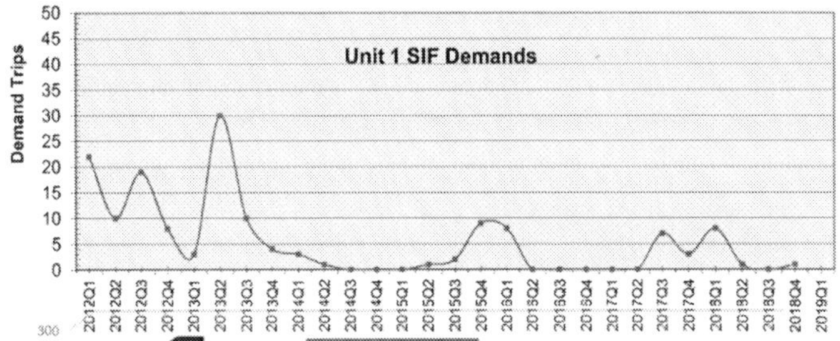

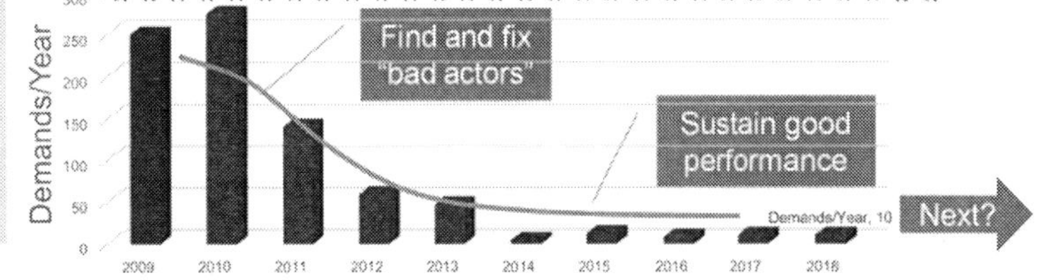

DUPONT Demands on "Last line of Defense" down <u>96%</u> over 10 yrs

10/16/2019 9

Case 2: Bypassing can add Risk

A different company recently identified the following situation:

- **Three identical production trains**
- **Each had a significant hazard protected by SIL3 Safety Function**
- **Design satisfied SIL3**
- **Demand rate As shown**
- **Bypass rate As shown**
- **See Fault Tree Diagram**

- Leadership of the Operating Company had no idea.
- Analysis brought proper resources to bear.

Demand Rate	
Historian Tag	**6 months data**
SIF-TrainA	2
SIF-TrainB	159
SIF-TrainC	15

- SIFs experiencing more than 1 or 2 real demands per year must be designed for "continuous demand mode"
- Why is Train B demand rate so much higher?

Bypass Rate	
Historian Tag	**6 months data**
HS-SIF-TrainA	72
HS-SIF-TrainB	70
HS-SIF-TrainC	58
Grand Total	200

- SIF is Bypassed ~130x/year for each of 3 Trains
- Admin Safeguard on the Bypass
- Error rate on Admin Safeguard?

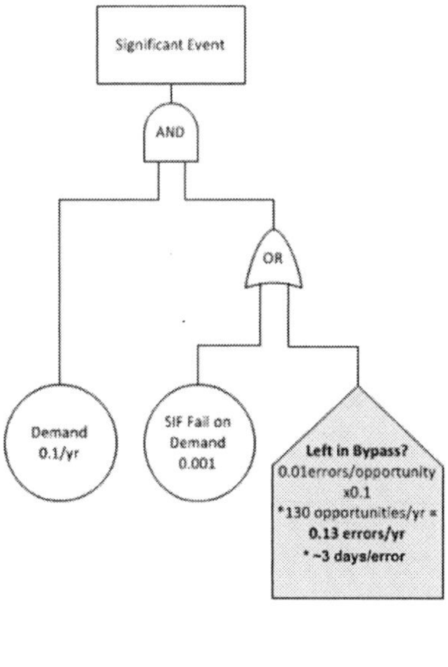

The Learnings

"In God we trust. All others must bring data."

– W. Edwards Deming

1. **Incorrect Use of SIFs (easy fixes)**
   - Operations allowing SIFs to act as regulatory controls and shutdowns
     - ➢ Trained operators designed alternative controls
   - Sequence operations used SIFs as regulatory stops and shutdowns
     - ➢ Redesigned sequences
2. **Some real issues identified (less easy)**
   - Common utilities caused multiple simultaneous trips
     - ➢ Improved utility control
   - Change in operations personnel cause some spikes
     - ➢ Improved operator training

DUPONT

The data was full of surprises. No one knew! 10/16/2019 11

The Future

- **First generation application is reaching end of life**
- **Piloting new application:**
  1. Event data for better timestamp resolution
  2. Connectivity with other data sources
  3. Improved analysis
  4. Improved reporting

DUPONT

10/16/2019 12

2019 EUROPEAN CONFERENCE ON PROCESS SAFETY & BIG DATA
OCTOBER 1-2, 2019
DECHEMA HOUSE, FRANKFURT AM MAIN, GERMANY
CCPS EPSC

Welcome from CCPS & EPSC!

- **Pol Hoorelbeke, Total, EPSC Board Member**
- **Willi Meier, DECHEMA, CCPS European Regional Manager**

2019 EUROPEAN CONFERENCE ON PROCESS SAFETY & BIG DATA
OCTOBER 1-2, 2019
DECHEMA HOUSE, FRANKFURT AM MAIN, GERMANY
CCPS EPSC

Thanks to Planning Committee Members:

1. Abdul Aldeeb - *Siemens*
2. Gus Carroll - *Centrica*
3. Ivan Castillo - *Dow Chemical*
4. Tony Downes - *Honeywell*
5. Arthur Groot - *DSM Resins & Functional Materials*
6. *Pol Hoorelbeke - TOTAL*
7. *Tijs Koerts - EPSC*
8. *Daniel Mottaghi – TOTAL*
9. Rendall Ricardo - Dow
10. Richard Pawulski - *Air Liquide*
11. Hans Schwarz – ProsafeX
12. Mike Snyder *DEKRA*

CCPS Staff:
Shakeel Kadri
CCPS Executive Director & CEO
Willi Meier
CCPS European Regional Manager
Jing Chen
CCPS Senior Engineering Specialist
Horst Massong
CCPS Europe Project Manager
Evan Flach
AIChE Associate Director, Sales

2019 EUROPEAN CONFERENCE
ON PROCESS SAFETY & BIG DATA

OCTOBER 1-2, 2019
DECHEMA HOUSE, FRANKFURT AM MAIN, GERMANY

CCPS EPSC

THANK YOU TO OUR CONFERENCE SPONSORS

GOLD

AcuTech
PROCESS RISK MANAGEMENT

SILVER

EMERSON. evision® SIEMENS Ingenuity for life

BRONZE

sphera™

2019 EUROPEAN CONFERENCE
ON PROCESS SAFETY & BIG DATA

OCTOBER 1-2, 2019
DECHEMA HOUSE, FRANKFURT AM MAIN, GERMANY

CCPS EPSC

THANK YOU TO OUR EXHIBITORS:

- AcuTech Consulting Group Inc
- CGE Risk Management Solutions
- Emerson
- eVision Industry Software
- Siemens Process & Safety Consulting
- Sphera
- Versatec Energy B.V.

Conference Date: October 13 – 14, 2020

DECHEMA House, Frankfurt AM Main, Germany

www.aiche.org/ccpseuropeanconference

Moment of Safety:

Applying Safety Analytics for Improving Organizational Risk Capability to Reduce Incident Frequency and Severity: From Heavy Industry to the Modern Tech World

Joe Stough

Oct 2, 2019

Outline of Presentation

- Introduction – history of experience
- Analyze Data
- Apply Analytics
  - Learning from Energy and Chemicals (latest benchmarking)
  - Learning from another universe (Waitr – food delivery app company)
- Questions
  - And a question for you...

Joe Stough & Process Safety Big Data

1995 → 2019

Joined Process Safety Family in '95
- Started Syntex, built MOC & PHA in '95
- Process Safety Family taking shape in '90's
- Old proprietary tech in '90's to new "open" technology approaching 2000
- Enterprise web-based in 2000's to cloud-based tech in 2010's
    - much more conducive for big data analytics
- Process Safety Big Data Family (PSBD) taking shape → now having annual family reunions @ CCPS-EPSC Big Data conference

This session: share experiences with tech, data, and analytics over this timeframe

Joe Stough - Family

Avid sports fan and outdoorsman from Louisiana
25 yrs software and analytics entrepreneur, founder of Syntex
Board of Directors – First Federal Bank and Ducks Unlimited
Educated at Univ of California, SB - Statistics grad degree

Stages of Evolution to Apply PS Big Data

Build Software			Deploy Globally		Analyze Data		Apply Analytics		
1995	1998	2002	2005	2008	2010	2013-2015	2016	2017-2018	2019
Founded Syntex	IMPACT Safety	IMPACT ERM (OE)	> 100 countries, > 20 languages	Multi-year Global data set	Sold Syntex to IHS, OE Analytics Research	Progressive (maturity) Index	Pragmatic Analytics	President → Waitr Excellence (WE)	Finish what you started

Risk Process Platform
- Incident investigations
- Audits, assessments, etc.
- PHA, MOC, etc.
- Preventive/corrective Action Items for ALL

Single Instance Enterprise Data
- Hundreds of total processes
- Tens of thousands of facilities (assets)
- > Million users

Statistical Analysis
- Collaborative with large global co's
- Hundreds of metrics in data set
- Iterated to find predictive index
- Analytics per risk process maturity

Affect Change with Analytics
- Pragmatic to engage leaders
- Readiness / maturity-based

The Focus of This Presentation

Start with End in Mind

ZERO Incidents

- If you had ZERO incidents and were doing the things that are proven mathematically to yield ZERO incidents, you wouldn't have a problem.
- You have an entire workforce doing things to prevent incidents: audits, assessments, observations, investigations, action items, etc...

THE question behind 25 years of work

- Given that some assets (in similar inherent risk categories) have such lower rates of incidents than others – i.e. the high performers – what are they doing differently to get so much closer to ZERO?
  - 2008-2015: Answer this question using data and analytics
  - 2015-Now: Apply the analytics to justify change and focus in areas of weakness

10/18/2019 5

Analyze Data

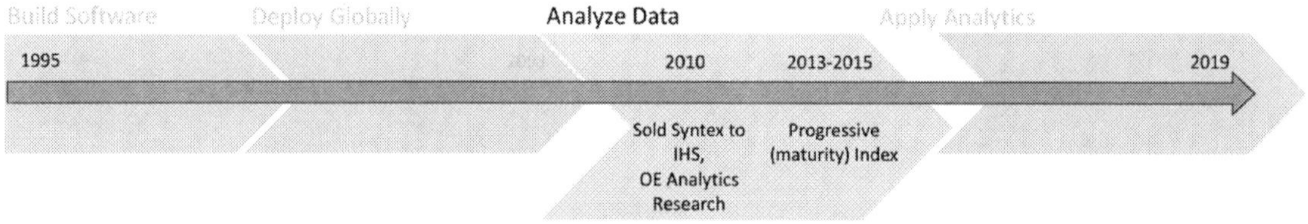

Statistical Analysis
- Collaborative with large global co's
- Hundreds of metrics in data set
- Iterated to find predictive index
- Analytics per risk process maturity

Best Achievement
- Strong mathematical, academic outcomes aligned with organizational maturity / readiness

Biggest Challenge
- Indexes difficult to translate pragmatically for practical implementation

10/18/2019 6

Since 2008, Research and Analytics

ARKEMA, CITGO, PHILLIPS 66, Chevron, KOCH INDUSTRIES INC., ExxonMobil, OMV, ADM, TOTAL (Chemicals Only), Shell, PPG, VALERO, lyondellbasell, ENERGY TRANSFER

Original (founding) participants in analytics research forum

Analytics Research: To Answer the Question...

What are the actionable, measurable factors that LEAD to below results?

LAGGING DATA → OVER A MILLION INCIDENTS

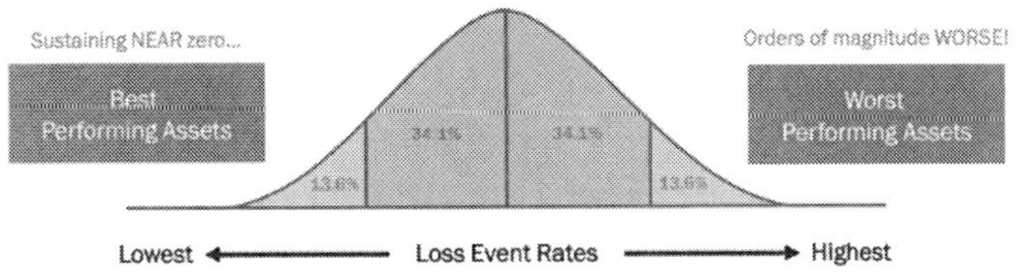

Workers Execute Processes to Strengthen Barriers

"A plant is not inherently safe, people make it safe"
Todd Conklin

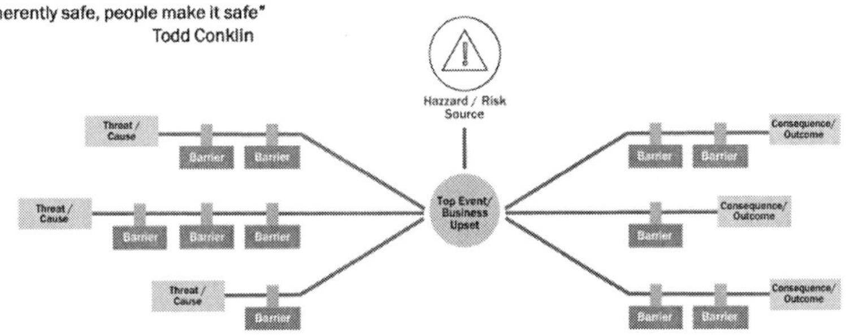

Management System (Risk) Processes:
Identify Gaps and Strengthen Barriers

- Near miss, incident investigation, and action item management
- Risk Assessment, PHA, HAZOP, LOPA
- Management of Change
- Audits and management system assessments
- Many, many others....

Risk Process Execution Data

THE DATA SCIENCE HYPOTHESIS AND ANALYTICS FINDINGS

The Hypothesis

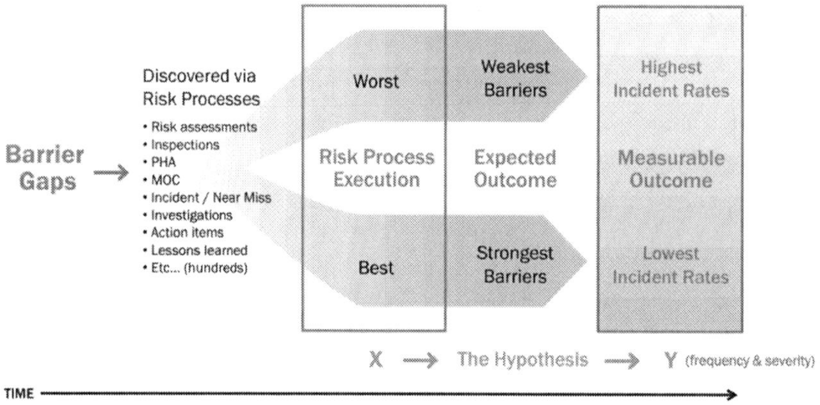

f(X) = factors of risk process execution BOTH most predictive of Y AND most practically actionable by operating units

9

THE DATA SCIENCE HYPOTHESIS AND ANALYTICS FINDINGS

2008-2013: SME's and Data Scientist Collaboration

Looking for f(X) = factors of risk process execution BOTH most predictive of Y AND most practically actionable by operating units

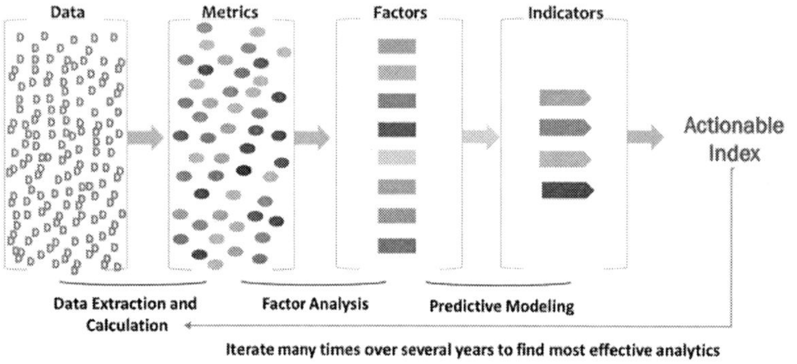

Analysis Results
- **2010 first index ~.47 R^2**
- **2013 best index >.7 R^2**

10

THE DATA SCIENCE HYPOTHESIS AND ANALYTICS FINDINGS

2014: After 5 Years of Analytics Initiatives...

Strongest R^2 for assets with highest risk process capability (learning-minded).

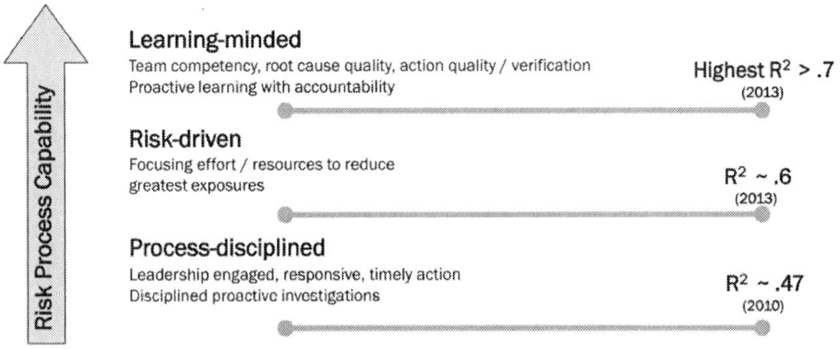

11

Apply Analytics

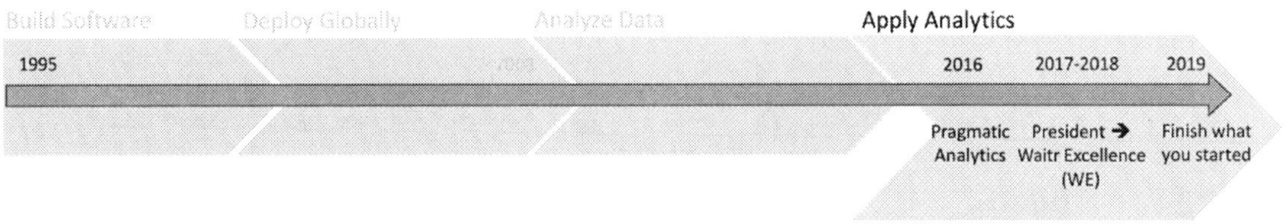

Affect Change with Analytics
- Pragmatic to engage leaders
- Readiness / maturity-based

Best Achievement
- Pragmatic roadmap for measuring / improving risk capability for any level of maturity / readiness

Biggest Challenge
- Adaptiveness of leadership either because of maturity-level or traditional resistance to applying analytics

10/18/2019 12

What We Learned From the Analytics

Practically speaking, what are they doing to get closer to ZERO?

- **Stage 1: Focus on Engagement and Discipline (Leadership / Management as Stewards)**
  - They take the visible lead in supporting risk processes (after much training and influence from the very top)
  - The **mean time to respond** to risk exposures is BOTH low (mean) and consistent (relative variation)
  - They integrate updates on material event investigations and action items into weekly management business reviews
    - Higher percent of material events investigated
    - Higher percent of high priority action items completed on-time
- **Stage 2: Focus on Materiality**
  - Because all of the above is in place, collecting a high volume of risk event data (investigations, PHA's, audits, etc.).
  - A **risk matrix** is implemented to quantify materiality of risk exposure "opportunities"
  - All above measures are statistically significantly stronger for high material than lower risk exposures
- **Stage 3: Focus on Quality**
  - Because all of the above is in place, leaders can focus on improving quality of response to most material exposures.
  - A **quality scoring method** is used to quantify quality of event reports, root causes, investigation teams, actions, etc...
  - Leaders are held accountable both to participate in peer quality reviews and to meet quality expectations

13

Using Pragmatic Language

2016: Moved to less academic, more practical language and metrics.

Evolving Areas of Focus

Expectations	Sample Measurements
Quality Leadership held accountable for quality of root cause and action items for material exposures.	Quality of root cause analysis for material exposures. Quality of action stewardship for material exposures.
Materiality Measurably shift resources and effort toward repairing barriers to higher material exposures	Difference in means between material and non-material events for process execution metrics.
Engagement and Discipline Establish consistency in execution of risk processes. Leadership responsiveness and action item stewardship.	Relative variation for consistency. Mean responses and action timeliness.

14

Most Prevalent Industry Effort: Materiality

The middle of the maturity bell curve across observed assets is "Materiality"

Evolving Areas of Focus

Expectations	Sample Measurements
Quality Leadership hold accountable for quality of root cause and action items for material	Quality of root cause analysis for material exposures. Quality of action stewardship for material exposures.
Materiality Measurably shift resources and effort toward repairing barriers to higher material exposures	Difference in means between material and non-material events for process execution metrics.
Engagement and Discipline Establish consistency in execution of risk processes. Leadership responsiveness and action item stewardship.	Relative variation for consistency. Mean responses and action timeliness.

15

2016: 12 Company Benchmarking

- 2015 Data
  - Total workhours: **920,884,650.28**
  - Total Safety Incidents: **19,213**
  - Total Event Action Items: **49,628**
- Safety Performance
  - TRIR (recordable incident rate)
    - ~ 6 co's > .5
    - ~ 6 co's < .5

      (using 200,000 as denominator)

Injury/Illness Severity	
Fatality	8
Lost / Rest Time	598
Recordable	1333
Non-Recordable	5334
Near Miss	11940

Event Type	
High Potential Incident	4,581
High Potential Near Miss	3,413
Incident	43,161
Near Miss	34,046

16

Example of Benchmarked Analytics

	Engaged		Disciplined		Materiality		Quality	
	NEAR MISS RATE	PROCESS ENGAGED	% ON TIME ACTIONS	% MITIGATED	% HIPO MITIGATED	RISK DRIVEN ACTION	MITIGATION OF HIPO RISK	ROOT CAUSE DISCOVERY
	219.0	80.70%	87.30%	0.30%	2.00%	52.00%	57.60%	8.20%
	22.3	71.70%	93.20%	50.00%	0.00%			
	511.0	84.30%	78.60%	4.00%	20.30%	52.20%	64.10%	28.80%
	12.7	80.30%	82.70%	33.10%	74.00%	62.10%	66.10%	74.40%
Your Co.	345.0	74.80%	85.80%	25.30%	14.30%	93.60%	68.30%	93.10%
	8.2	47.10%	77.50%	57.70%	70.10%	57.80%	58.50%	76.90%
	68.7	59.50%	74.20%	2.80%	0.00%			
	212.0	67.80%	65.90%	31.20%	41.30%	29.20%	10.80%	61.50%
	118.0	81.30%	57.70%	11.00%	13.00%	11.90%	27.70%	34.10%
	45.3	66.30%	74.50%	38.00%	0.00%			
	22.8	47.40%	53.20%	0.30%	1.00%	9.60%	63.40%	7.80%
	93.8	66.10%	75.40%	29.30%	0.00%			

Shaded scores reflect areas where "process maturity" is insufficient.

WHAT DOES EXCELLENCE LOOK LIKE?

A Materiality-focused Organization

13 operating assets within a single division of a high-performing company

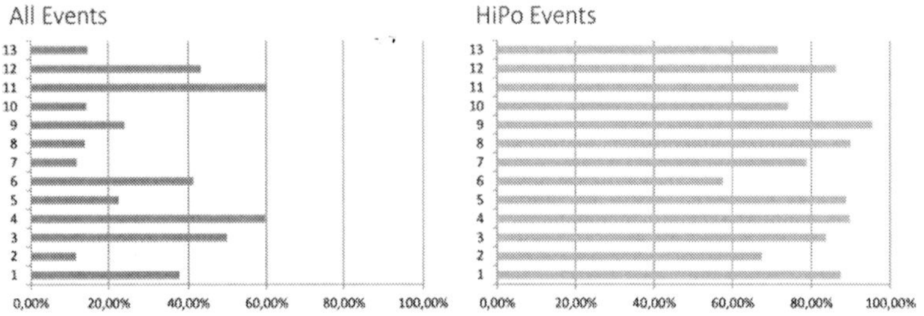

31.18%	μ	80.60%
18%	σ	11%
59%	σ/μ	13%

Substantial difference in performance and consistency.
Reflects a risk-driven process and a focus on Materiality.

Learnings from 25 Years in the Family…

- There are assets in heavy industry performing much closer to ZERO
  - Even in same company compared to similar asset types
  - Definitely across companies
- Excellence requires discipline with focus on materiality and quality
  - But YOU (many business leaders) already knew that
  - However why are so few of the assets really strong in these areas?
- Next steps to increase number of excellent assets
  - Identify behaviors (metrics) relevant for next step in risk process capability
  - Educate / engage top business leaders as committed stewards
  - Hold asset leaders accountable to improve metrics that yield excellence

Applying Learnings to a Digital Tech Company

Waitr – a food delivery app business in USA. (similar to Uber Eats, Delivery Hero, etc.)

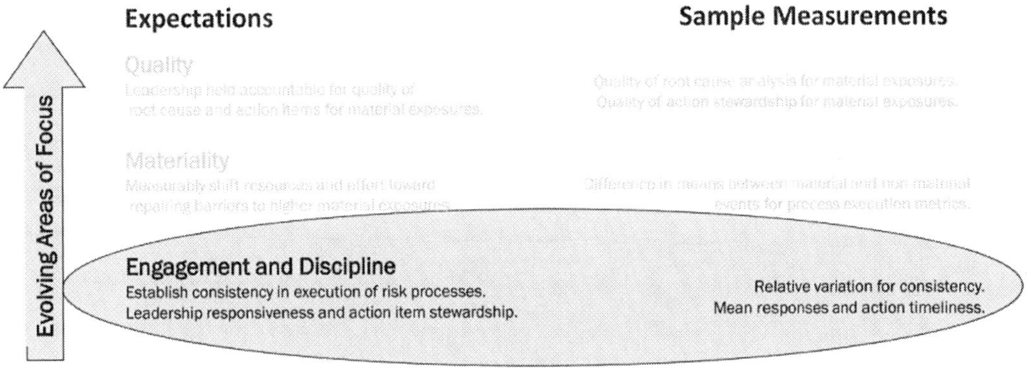

WAITR Excellence

- Leadership Stewardship
  - Me as President of Public Company
  - Key capabilities not "natural" for managers
- Key Terms
  - Something COULD go wrong = Risk
  - Something DID go wrong = Incident
- WE is an Operational Excellence Management System to strengthen the protections from the things that could go wrong at WAITR
  - Delivery service quality (late deliveries)
  - Restaurant onboarding timeliness (throughput)
  - Customer service responsiveness
  - Etc...

Refund Deviations: New Incident Process 2019

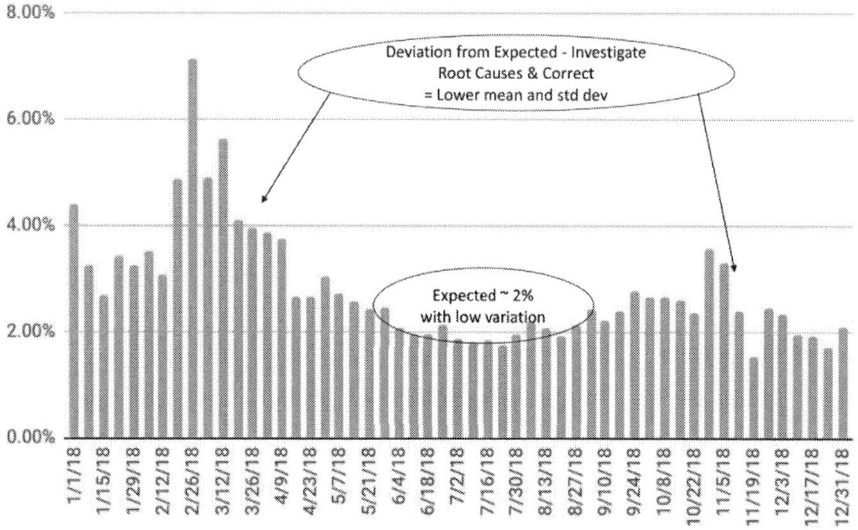

WE Risk Reduction Process

5-Why: Why did (could) something go wrong?

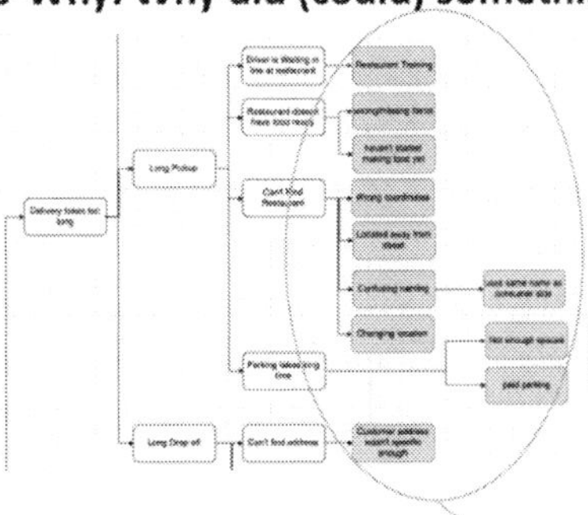

Root Causes

5-Why

Operating Stability of Market
➔ Strength of Protections

Market Risk Assessment

Implement Protections

Corrective / Preventive Actions

Data & Metrics

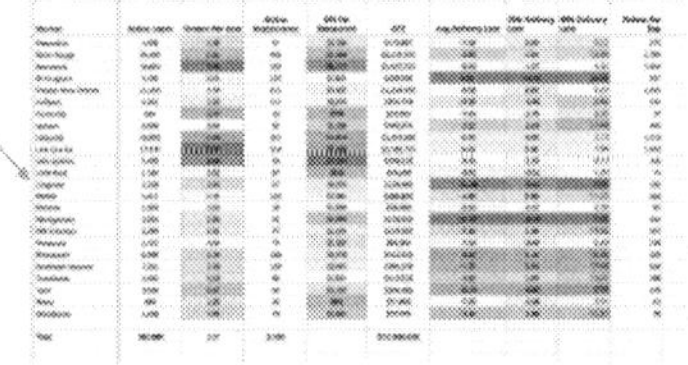

Flash Reports / KPI Dashboards

Key Learning from WE

- Applying analytics and risk process practices in low maturity Waitr
  - Young employees, new to such processes
  - Required very slow start to incubate understanding of the need
  - Data was the biggest catalyst for engagement
  - And all this with the President as most passionate supporter
- Behaviors that yield excellence is "unnatural" for humans
  - Not natural to openly invite deep dive into most egregious problems
  - Not natural to get to the true root cause
  - Not natural to skip over the easy problems to focus on the highest impact

Questions?

Question for you: what is the greatest current vacuum?

User friendly software platform that delivers BOTH deep risk process capability AND rich data sets across the enterprise

-OR-

Analytics solutions that leverage your existing deep and rich enterprise data sets to deliver impactful performance metrics in the business

10/18/2019 25

Stages of Evolution to Apply PS Big Data

CCPS EPSC

Build Software | Deploy Globally | Analyze Data | Apply Analytics

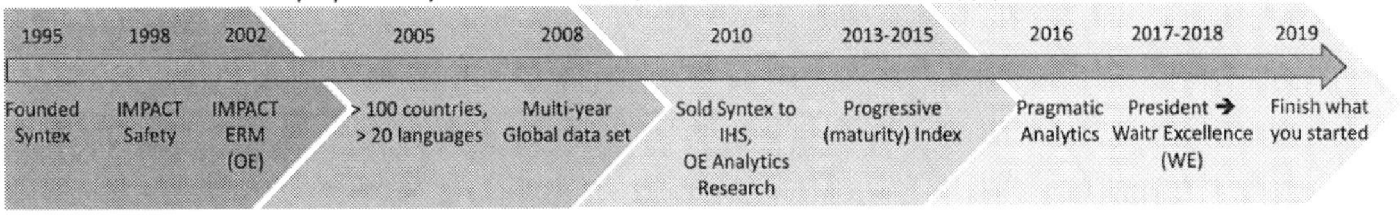

Risk Process Platform
- Incident investigations
- Audits, assessments, etc.
- PHA, MOC, etc.
- Preventive/corrective Action Items for ALL

Single Instance Enterprise Data
- Hundreds of total processes
- Tens of thousands of facilities (assets)
- > Million users

Statistical Analysis
- Collaborative with large global co's
- Hundreds of metrics in data set
- Iterated to find predictive index
- Analytics per risk process maturity

Affect Change with Analytics
- Pragmatic to engage leaders
- Readiness / maturity-based

The Focus of This Presentation

CCPS EPSC

Panel Discussion

Tony, the above slide is from my keynote. That is the context which the audience will have per the same timeline in my view of the state of the digital solutions in the market....

25 Years of Technology Lessons

CCPS EPSC

Build Software | Deploy Globally | Analyze Data | Apply Analytics

1995 | 2002 | 2008 | 2015 | 2019

- Open, enterprise systems begin to enable big data availability (seeds for PSBD Family)
- Tech options are either user friendly without sophisticated capability or sophisticated without user friendliness

- Open, enterprise systems create boom in data availability (PSBD Family is born)
- But analytics-driven solutions are slow to mature partly because senior leaders are reluctant to adopt sophisticated analytics from these systems

- Mobile, cloud-based systems have potential to combine data availability and user experience
- But the new solutions must strike balance between easy-to-use and sophisticated risk process demands
- THEN the next wave of truly analytics-driven solutions will be here!
- PSBD Family is Happy ⍰ ⍰ ⍰

10/18/2019 3

sphera™
Sphera is the largest global provider of Integrated Risk Management software and information services with a focus on:
Environmental Performance
Operational Risk
Product Stewardship
Mission: to create a safer, more sustainable & productive world
Simon Jones
Director, Digital Solutions Group
Simon.Jones@sphera.com

2019 PROCESS SAFETY SURVEY: KEY FINDINGS
Good Intentions
84%
Safety is a corporate value supported by president/CEO
70%
Have a defined roadmap & goals for PSM
63%
Well-defined leading & lagging safety indicators
But Falling Short on Meaningful Progress
40%
say organizations proactively manage process safety
77%
gaps between process safety intent and what actually happens
77%
risk increases between periodic process safety review periods

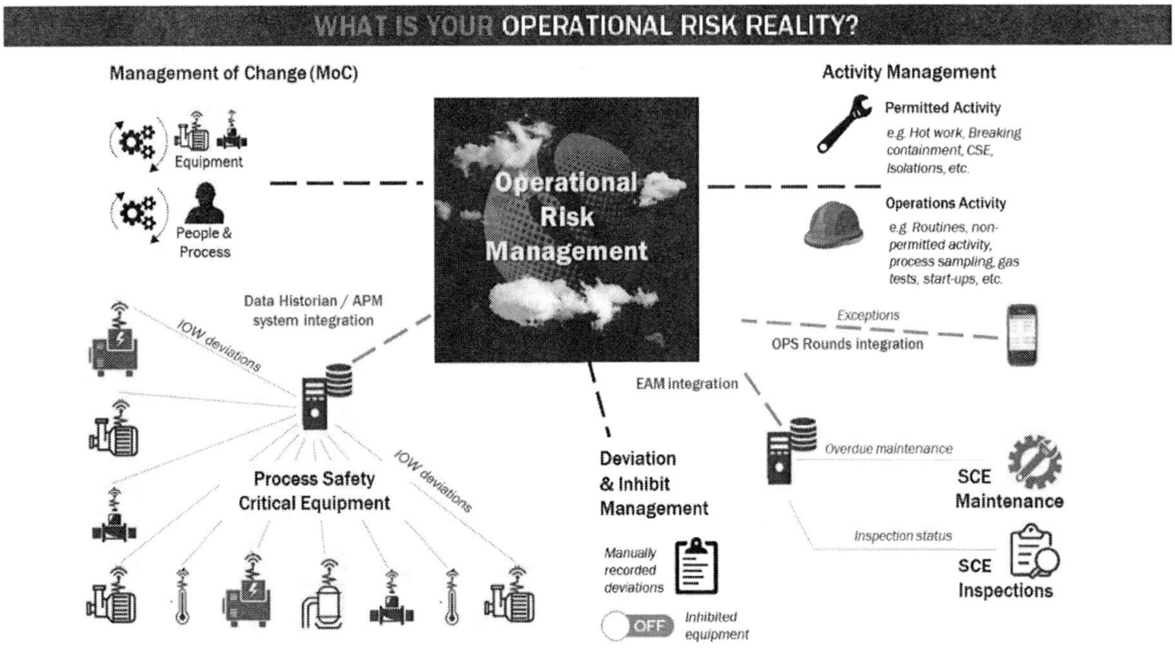
WHAT IS YOUR OPERATIONAL RISK REALITY?
Management of Change (MoC)
Equipment
People & Process
Activity Management
Permitted Activity
e.g. Hot work, Breaking containment, CSE, Isolations, etc.
Operations Activity
e.g. Routines, non-permitted activity, process sampling, gas tests, start-ups, etc.
Operational Risk Management
Data Historian / APM system integration
IOW deviations
IOW deviations
Process Safety Critical Equipment
Exceptions
OPS Rounds integration
EAM integration
Deviation & Inhibit Management
Manually recorded deviations
OFF
Inhibited equipment
Overdue maintenance
SCE Maintenance
Inspection status
SCE Inspections

DNV·GL

OIL & GAS

MyQRA: Bringing QRA studies into the digital age
&
Cloud QRA: Developing Digital Twins for Offshore Installations

2019 European Conference on Process Safety & Big Data

Olivier Baldan & Zoe Wattis

02 October 2019

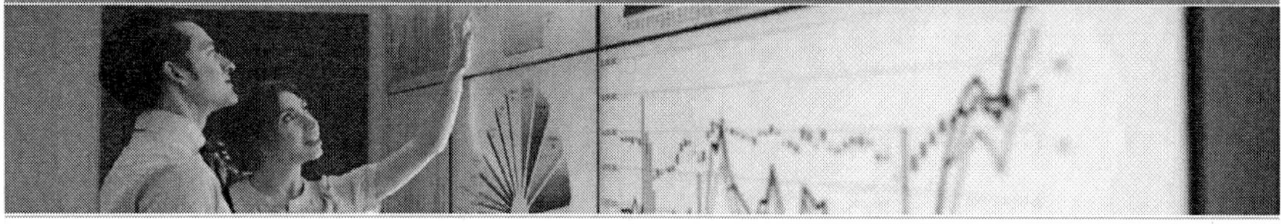

What are the challenges and issues with current QRA practices?

Accessing information from QRA studies

QRA reports offer certain information at a fixed point in time.

QRA models become more complex generating an increasing amount of data which is not easily accessible.

QRA data comes often too late to provide input at concept or design stage.

Understanding and visualization of risk

Professionals need a better understanding of hazards and the key risk drivers to make day-to-day decisions.

Safety-critical choices are sometimes made based on an outdated risk picture.

Standard QRA`s are delivered in data formats that are not easy to understand.

Processing and communication of results

Processing and communication of results require much resources and have a big potential for efficiency gains.

QRA data are difficult to hand over in the transition between CAPEX and OPEX phase

Sensitivity analyses take time and often the results are first available after decisions were made

DRAFT

DNV·GL
MyQRA & Cloud QRA
Efficient QRA and Online sensitivities
• Online Sensitivities
• Instant digital access
• More efficient updates
Improved risk communication and enhanced visualizations
• Improved communication across multiple stakeholders in a project or asset
• 3D Visualization
• Map Visualization
Interactive risk results
• Unlocks detailed information from QRA
• Filtering, drill-down functionalities powered by Power BI reports

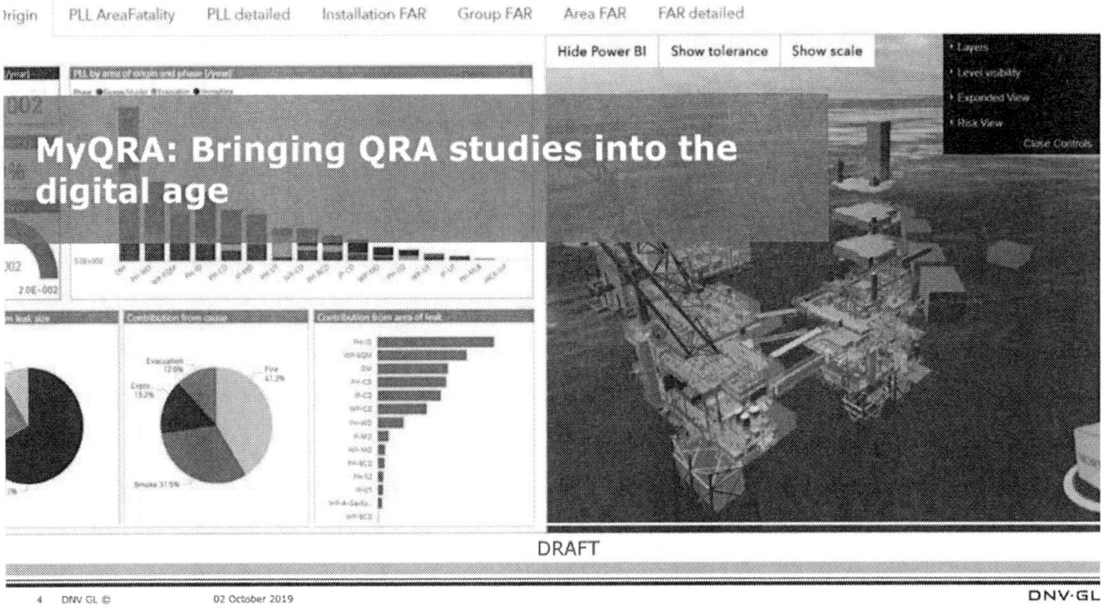
PLL AreaFatality
PLL detailed
Installation FAR
Group FAR
Area FAR
FAR detailed
Hide Power BI
Show tolerance
Show scale
MyQRA: Bringing QRA studies into the digital age
DRAFT
4 DNV GL ©
02 October 2019
DNV·GL

Results available on interactive map

- Display risk contours
- Display all results of consequence modelling
  - Outcome (Jet fire, pool fire, flash fire, …)
  - Leak size (small, medium, large)
  - Weather (Wind speed and stability)
  - Direction (North, South, East, West, envelope)

DRAFT

The user can explore the input of a risk model

- Improved communication and transparency
- Visualize input such as
  - Population
  - Ignition sources
  - Obstruction areas
- Toolbox with detailed information

DRAFT

The dashboard allows the user to drill down and filter to find the information he needs and visualize it in 3D models

- Interactive Power BI reports for all risk metrics (PLL, LSIR, FN, Escalation Frequency, Release Frequency, IRPA, etc.)
  - Reports are customizable
  - Comparison of QRA versions side by side
- 3D visualization

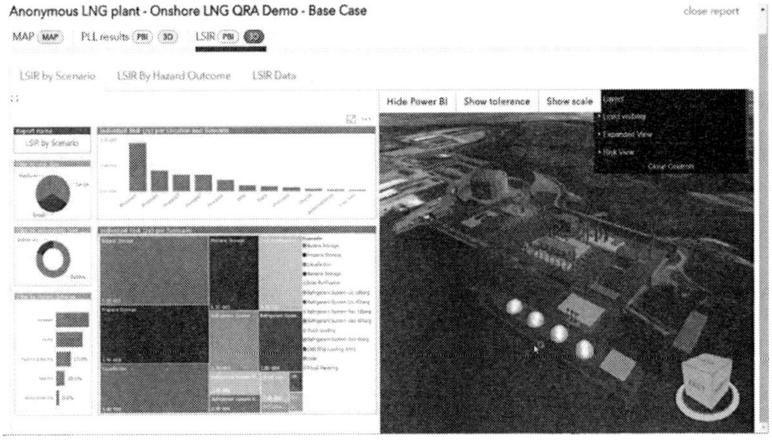

DRAFT

Detailed results and output can be shared with the user

DRAFT

Simple sensitivity analyses can be performed online

- A sensitivity on leak frequency can be performed
- The list of equipment for each segment can be modified to add or remove equipment, modifying the number of potential leak sources
- The resulting leak frequency is recalculated as well as the resulting risk and can be compared to the base case

- Other parameters are available such as:
  - Personnel
  - Wind rose
  - Ignition sources
  - ...

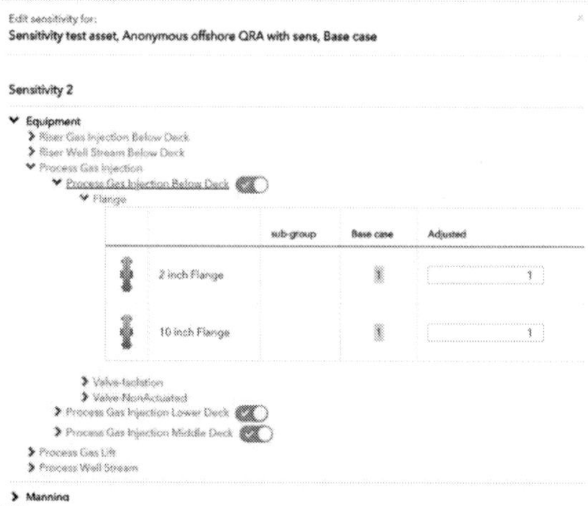

DRAFT

DRAFT

Objective

- To demonstrate how the cloud based tool CloudQRA can turn the traditional static Quantitative Risk Assessment (QRA) process into a "living QRA" which can be updated throughout a facility's lifecycle

CloudQRA - Data Capture

Data for a QRA is stored digitally, either in 'intelligent' systems or as scanned images of paper documents. CloudQRA:

- Uses embedded smart data to automatically define key information
- Uses tools to add information layers onto plot plans by "drag and drop" process
- Stores multiple sources of data in a central parameter repository
- Uses structured hierarchies to link platform and process information
- Allows inheritance of information which limits duplication and tracks changes

DRAFT

Automated Data Extraction

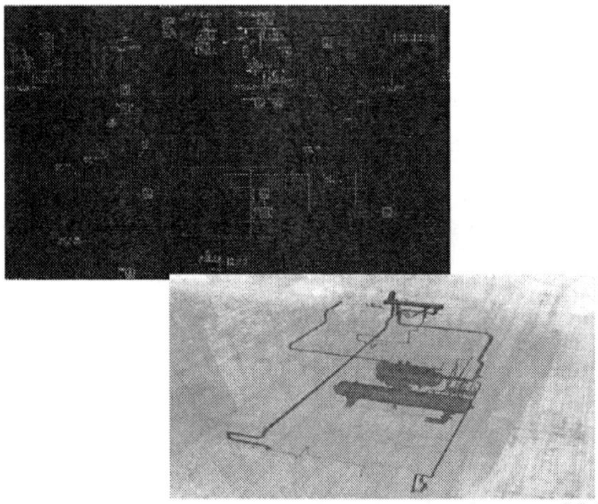

- Smart data tags in PFDs and P&IDs can be used in conjunction with sectionalisation
- Additional QRA data can be used to link P&IDs to the Heat Mass Balance and 3D CAD model
- Information produced includes:
  - Automatic parts count
  - Leak frequencies attributed to appropriate decks/modules
  - Release scenarios
  - Piping volumes
  - 3D models for use by consequence models

DRAFT

Graphical Data Input

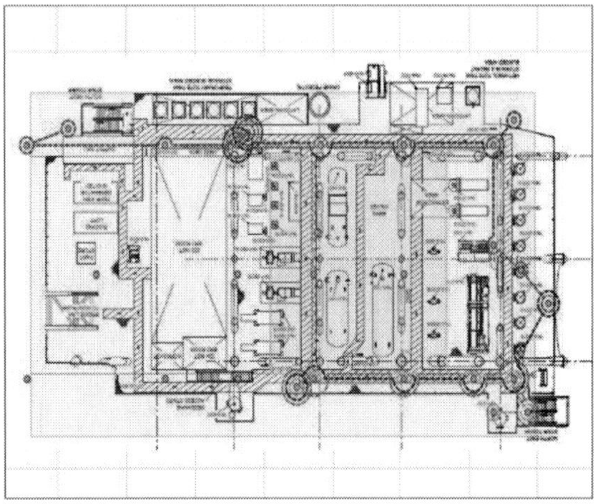

- Addition of information in layers to a graphical data source such as engineering drawings.
- Visual means of identifying aspects within model which can be tagged with further information
- Data capture tables can be used to store platform-specific data from spreadsheets
- Simplifies the process of defining co-ordinates for use in models

DRAFT

CloudQRA – Modelling and Results

Input data in conjunction with validated consequence and risk models running on the cloud stores results in a central database. Results can be extracted in a variety of formats including:

- Web/remote access
- Interactive dashboards and data interrogation
- Use of 2D and 3D visuals to display results
- Animation and potential for virtual reality simulation
- Ability for operators to adjust input data, rerun models and view results

DRAFT

Visual Presentation – Interactive Dashboards

- Aggregated metrics can be used to "drill down" into detailed results to provide insights not available in a static report
- Dashboard software can be linked directly to the Data Hub
- Output can be customised to develop alternative output templates depending on the requirements of the project

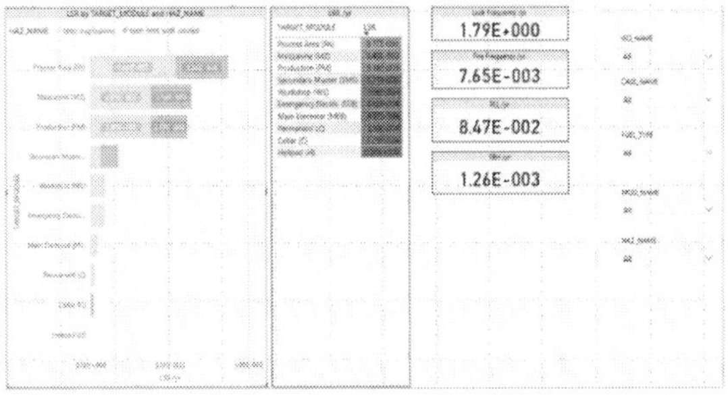

DRAFT

Visual Presentation – 2D Visuals

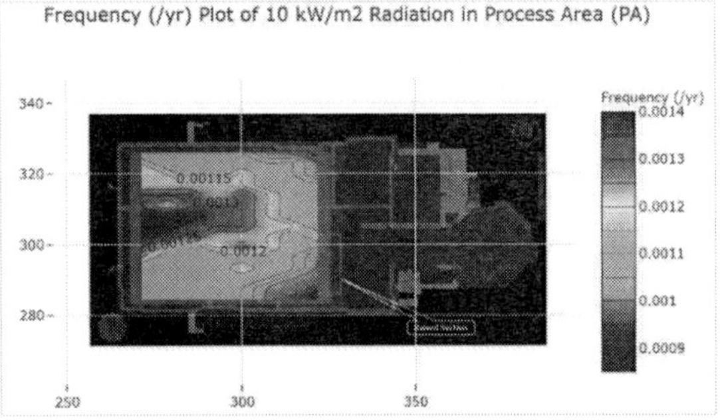

- Web browser tools can be used for visual representation of results
- Contour plots can be written to take slices of data from the database and viewed via a standard internet browser
- Presenting QRA data in tabular format allows results to be easily exported to spreadsheet

DRAFT

Visual Presentation – 3D Visuals

- Results can be layered on to 3D model build:
  - Semi-transparent coloured blocks illustrate areas with different risk levels
  - Red dots showing monitor points
  - Green dots showing release locations
  - Blue lines representing escape routes

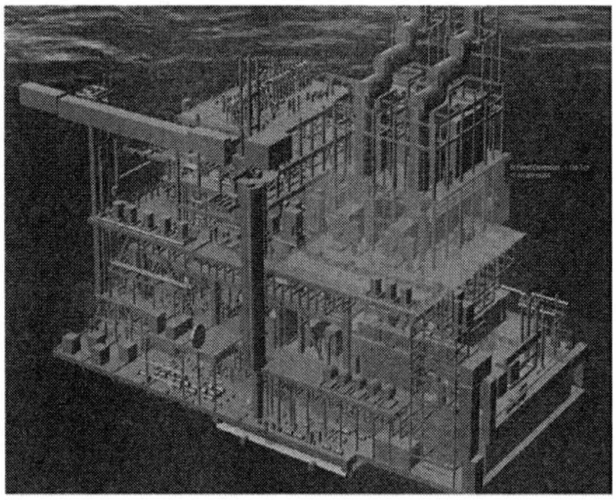

DRAFT

And Finally...

- Centralised data allows improved efficiency and easier comparison of data trends
- Calculation engines reside separately from data and interface providing flexibility if new models are introduced
- Hosting the data on the cloud allows multiple users to access information and secure user web access
- The cloud provides processor time in a scalable manner allowing models to be run simultaneously across multiple cores and sensitivities to be quickly generated
- Complex graphical models can be viewed in standard web browsers or interactive dashboards allowing easier extraction of data

DRAFT

Thank you for listening

Online demo of MyQRA is available from our digital marketplace on Veracity.com

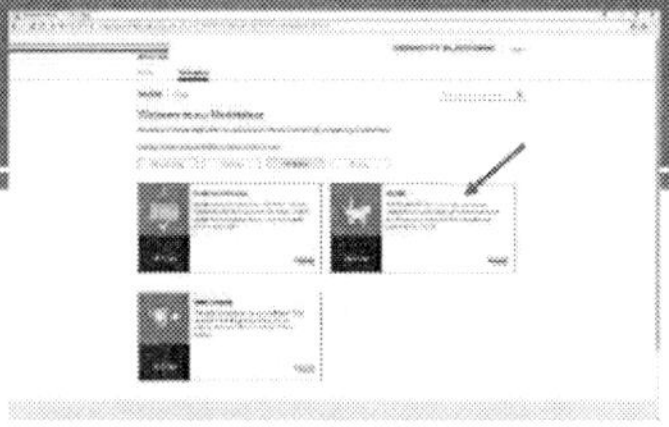

Zoe.wattis@dnvgl.com

Olivier.baldan@dnvgl.com

www.dnvgl.com

SAFER, SMARTER, GREENER

DRAFT

A Holistic Approach for Managing Past, Present and Future of Operational Process Safety

Steve J Elliott
Senior Director - Process Safety
Oct 2019

Life Is On | Schneider Electric

Steve J Elliott
Senior Director – Process Safety
+44 7785 371642
steve.j.elliott@se.com
LinkedIn: https://www.linkedin.com/in/steve-elliott-7148473/

Life Is On | Schneider Electric

Are you ready for exponential change?

Gerd Leonhard
[Futurist]

https://www.youtube.com/watch?v=ystdF6jN7hc

Life Is On | Schneider Electric

What does IIoT mean to you?

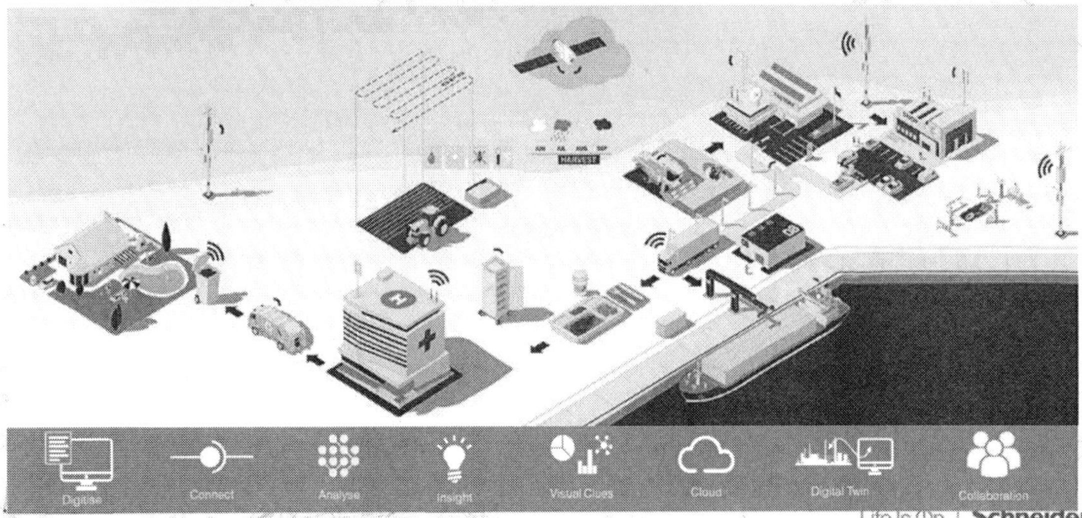

Life Is On | Schneider Electric

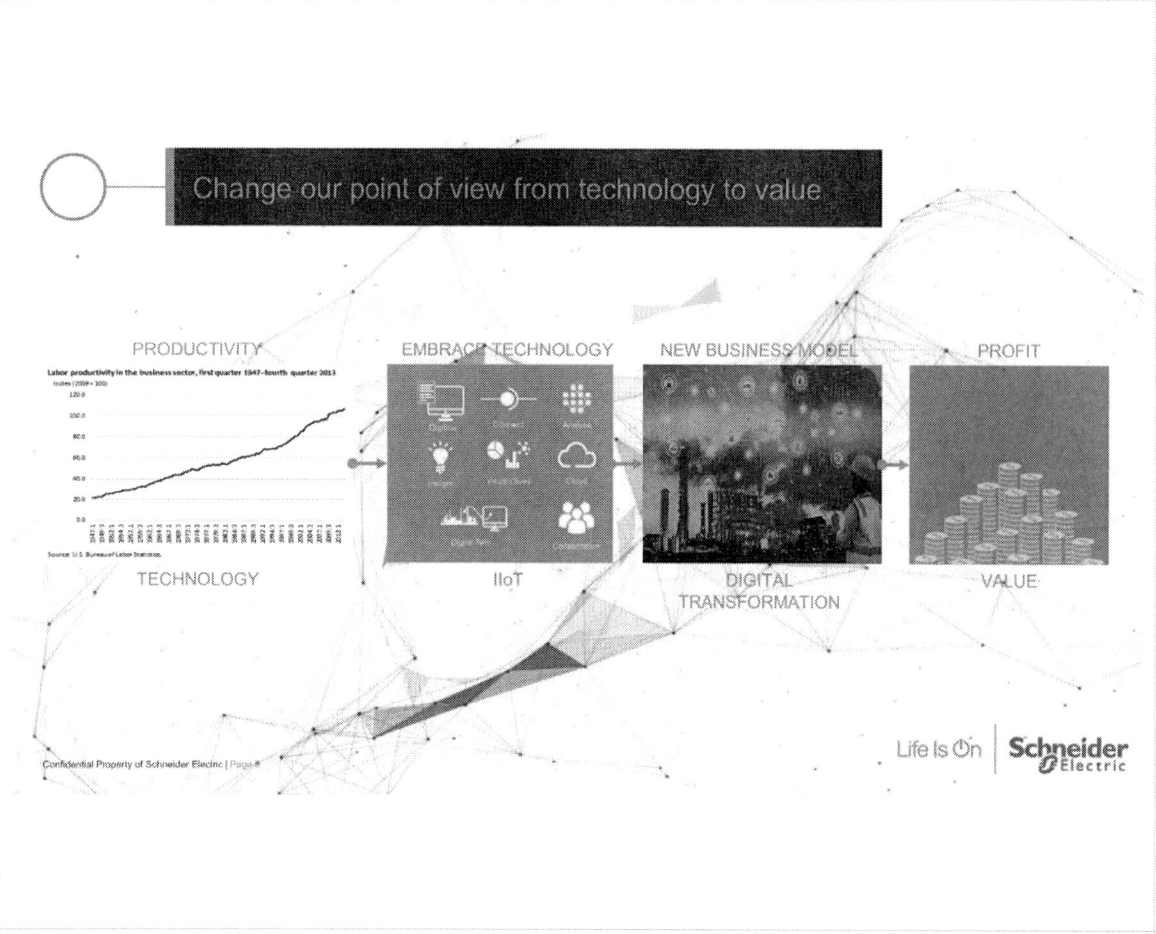
Change our point of view from technology to value
PRODUCTIVITY
Labor productivity in the business sector, first quarter 1947–fourth quarter 2013
Source: U.S. Bureau of Labor Statistics.
TECHNOLOGY
EMBRACE TECHNOLOGY
IIoT
NEW BUSINESS MODEL
DIGITAL TRANSFORMATION
PROFIT
VALUE
Confidential Property of Schneider Electric | Page
Life Is On
Schneider Electric

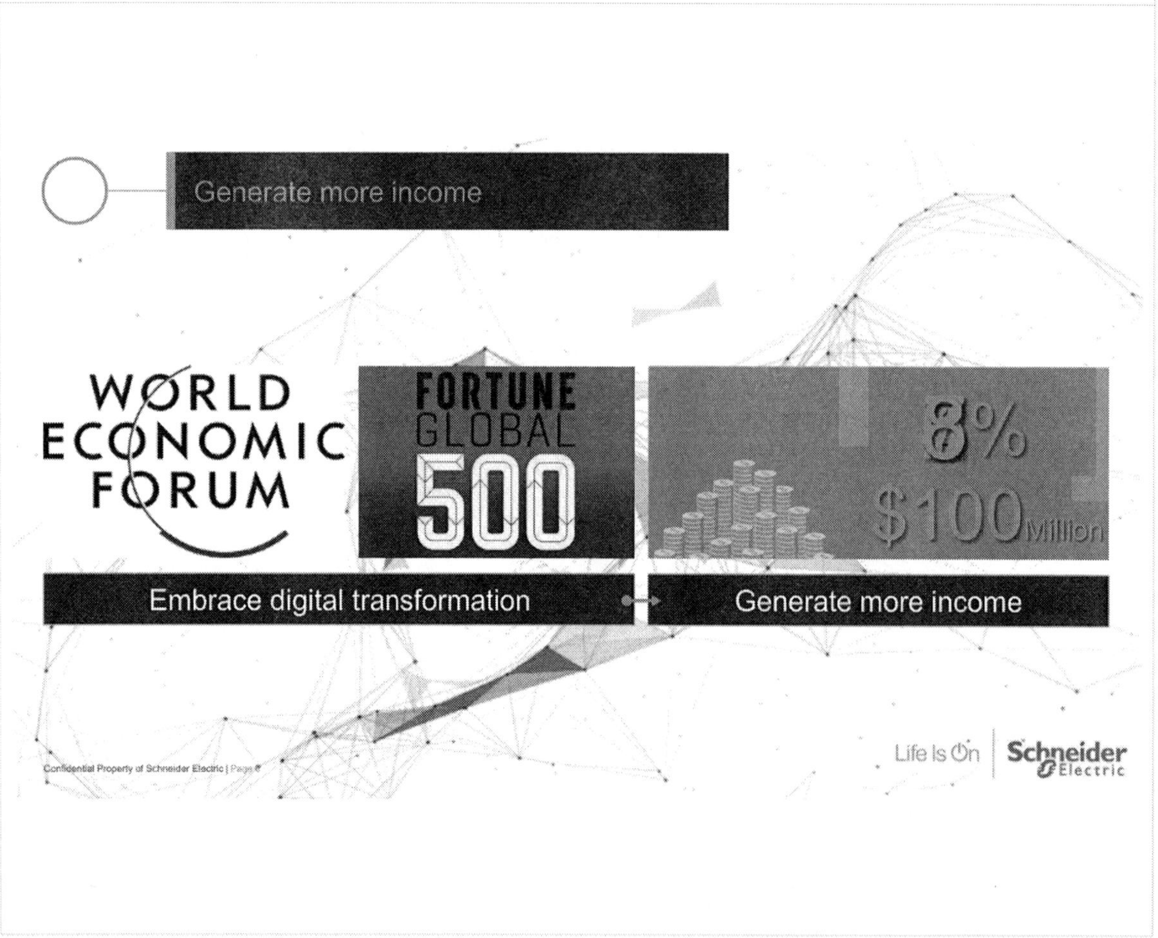
Generate more income
WORLD ECONOMIC FORUM
FORTUNE GLOBAL 500
8%
$100 Million
Embrace digital transformation
Generate more income
Confidential Property of Schneider Electric | Page
Life Is On
Schneider Electric

"Time is the enemy"
Confidential Property of Schneider Electric | Page

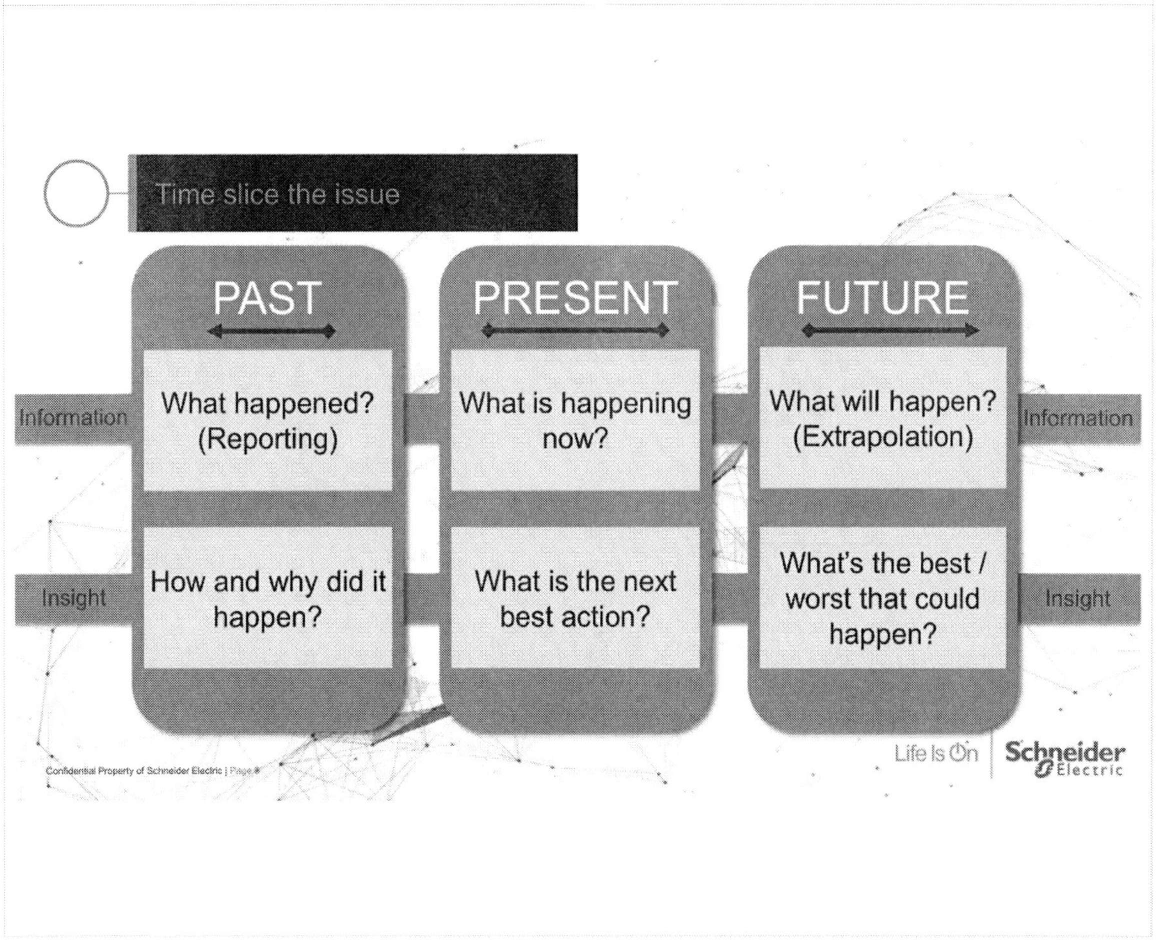
Time slice the issue
PAST
PRESENT
FUTURE
Information
What happened? (Reporting)
What is happening now?
What will happen? (Extrapolation)
Information
Insight
How and why did it happen?
What is the next best action?
What's the best / worst that could happen?
Insight
Confidential Property of Schneider Electric | Page
Life Is On
Schneider Electric

"Whiteboard moment"
Confidential Property of Schneider Electric

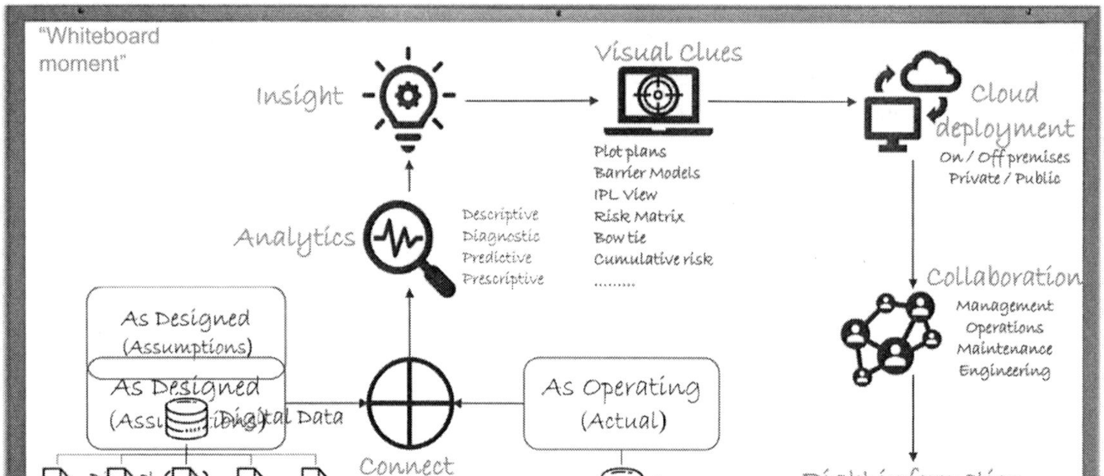
"Whiteboard moment"
Insight
Visual Clues
Plot plans
Barrier Models
IPL View
Risk Matrix
Bow tie
Cumulative risk
Cloud deployment
On / Off premises
Private / Public
Analytics
Descriptive
Diagnostic
Predictive
Prescriptive
Collaboration
Management
Operations
Maintenance
Engineering
As Designed
(Assumptions)
Digital Data
Connect
As Operating
(Actual)
Process
Data
Right information
Timely decisions
Improve control of
hazardous events

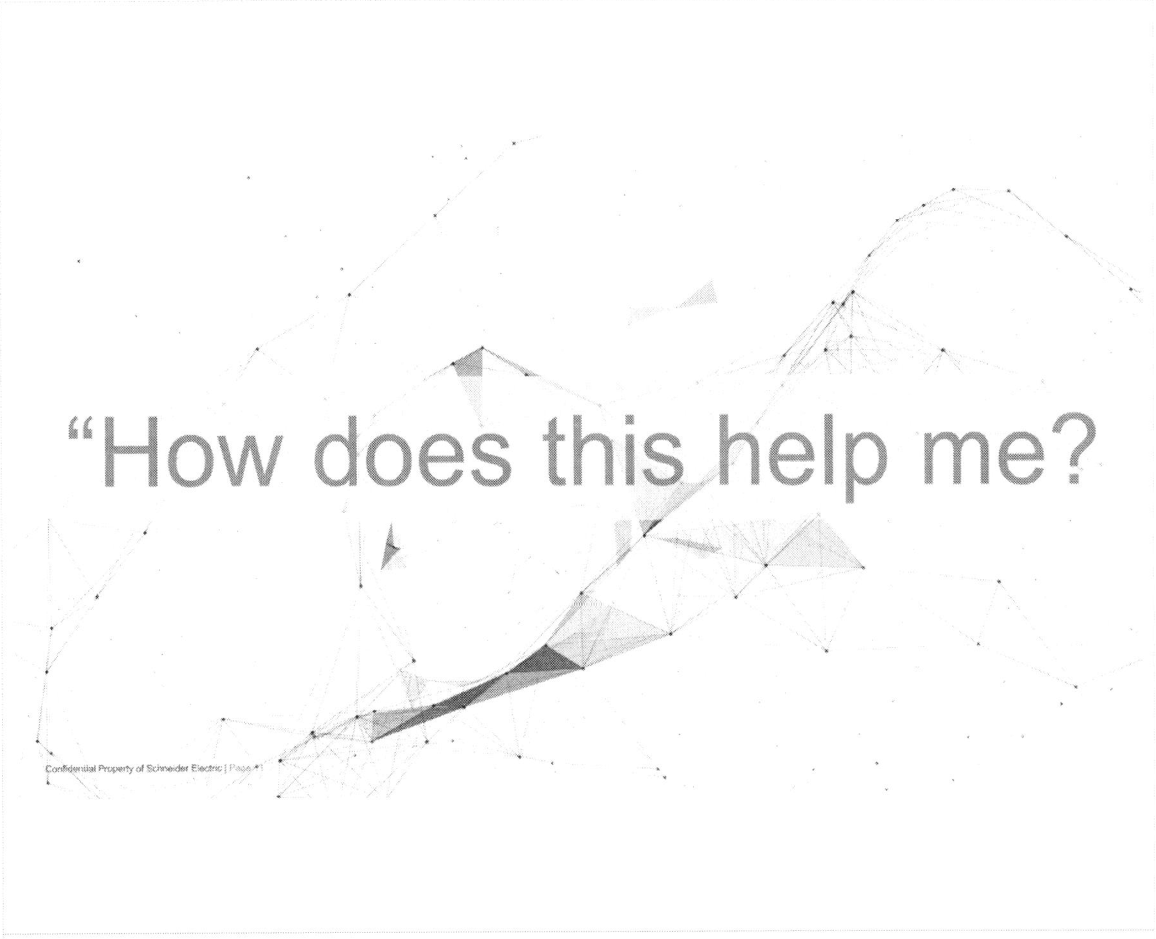
"How does this help me?
Confidential Property of Schneider Electric | Page #1

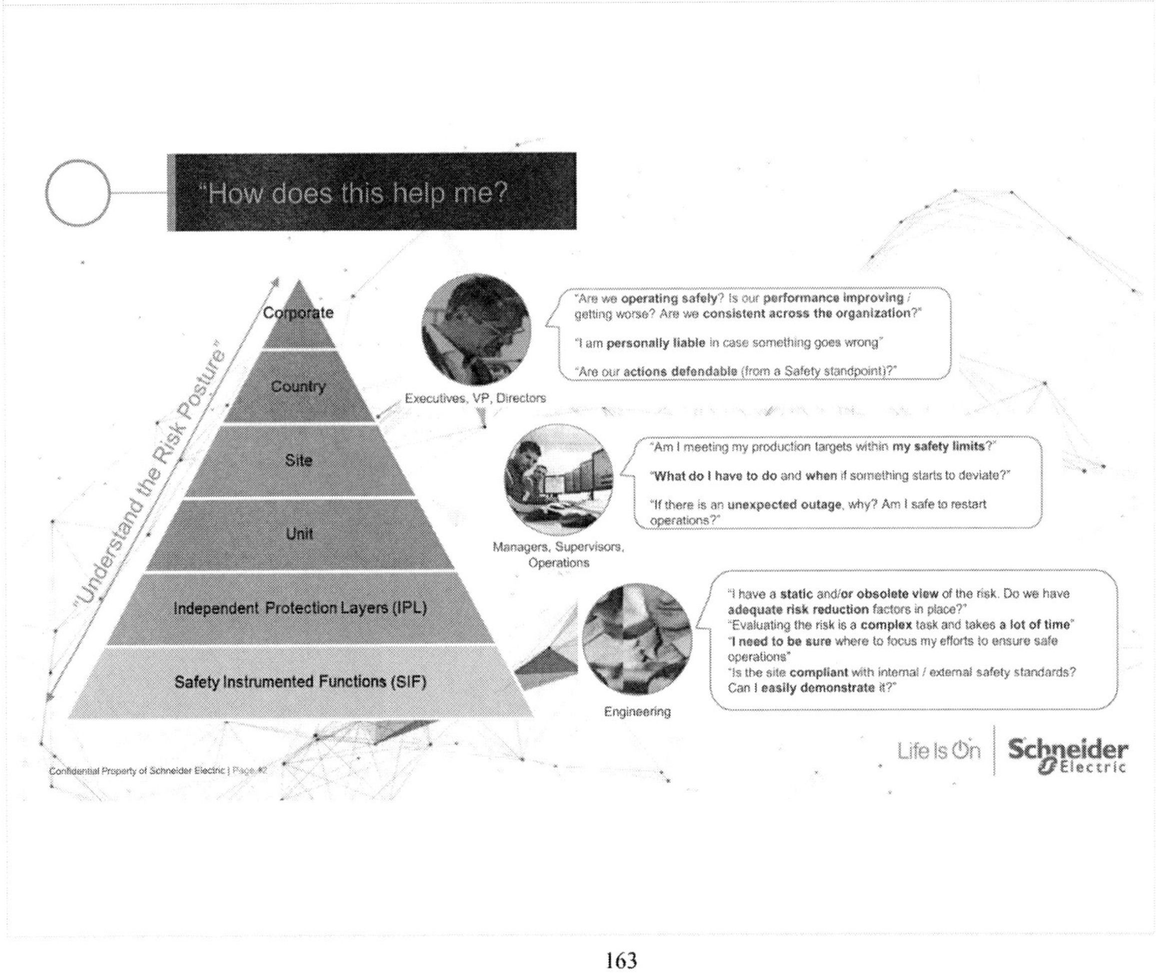
"How does this help me?
"Understand the Risk Posture"
Corporate
Country
Site
Unit
Independent Protection Layers (IPL)
Safety Instrumented Functions (SIF)
"Are we operating safely? Is our performance improving / getting worse? Are we consistent across the organization?"
"I am personally liable in case something goes wrong"
"Are our actions defendable (from a Safety standpoint)?"
Executives, VP, Directors
"Am I meeting my production targets within my safety limits?"
"What do I have to do and when if something starts to deviate?"
"If there is an unexpected outage, why? Am I safe to restart operations?"
Managers, Supervisors, Operations
"I have a static and/or obsolete view of the risk. Do we have adequate risk reduction factors in place?"
"Evaluating the risk is a complex task and takes a lot of time"
"I need to be sure where to focus my efforts to ensure safe operations"
"Is the site compliant with internal / external safety standards? Can I easily demonstrate it?"
Engineering
Life Is On
Schneider Electric
Confidential Property of Schneider Electric | Page #2

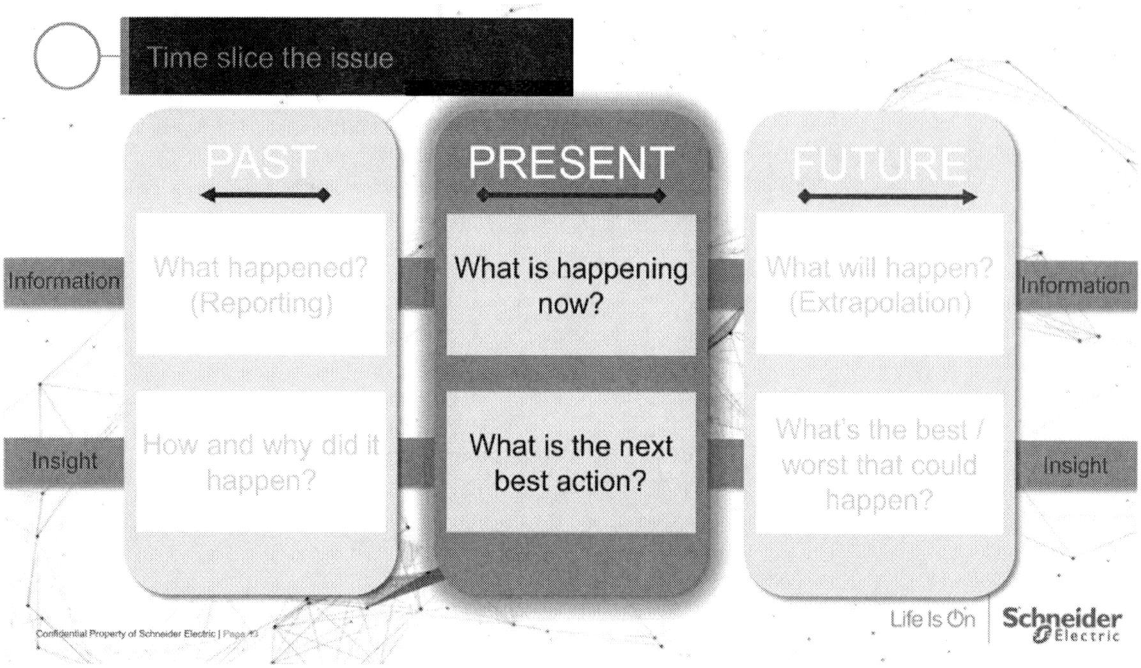
Time slice the issue
PAST
PRESENT
FUTURE
Information
What happened? (Reporting)
What is happening now?
What will happen? (Extrapolation)
Information
Insight
How and why did it happen?
What is the next best action?
What's the best / worst that could happen?
Insight
Confidential Property of Schneider Electric |
Life Is On
Schneider Electric

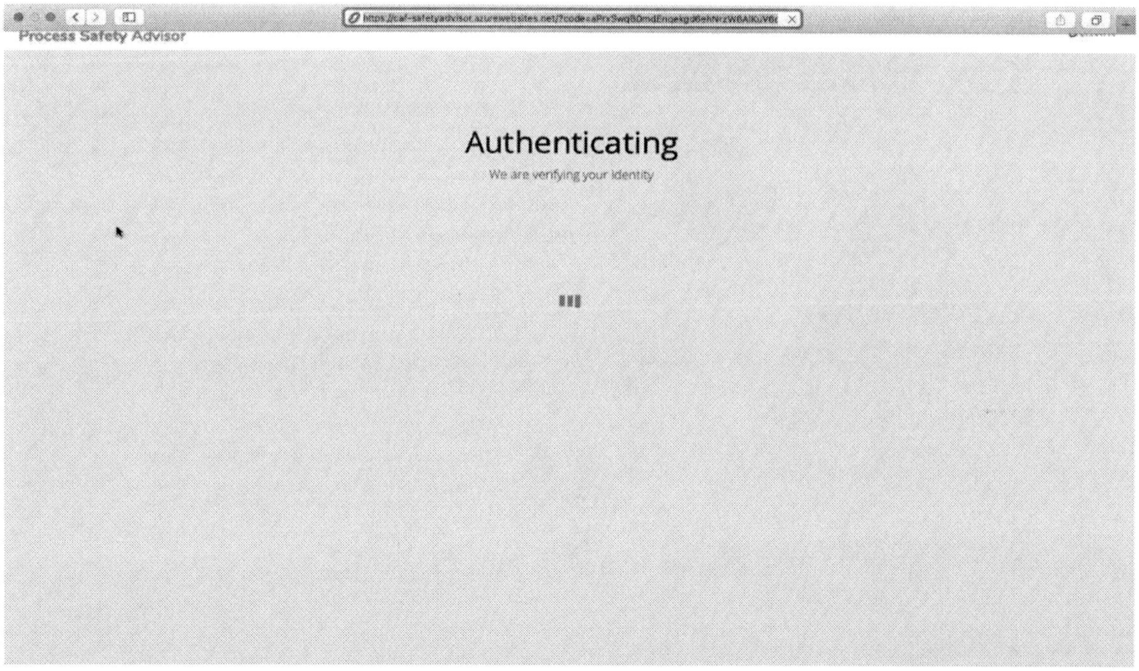
Process Safety Advisor
Authenticating
We are verifying your identity

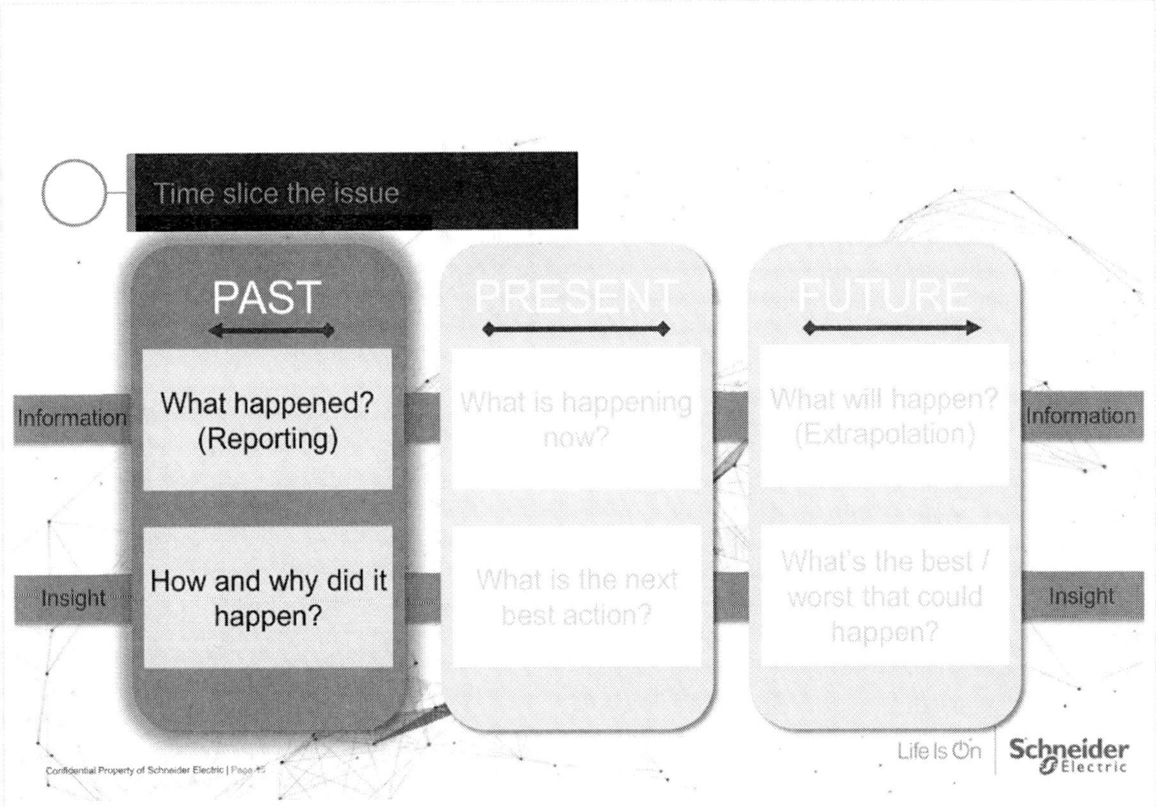
Time slice the issue
PAST
PRESENT
FUTURE
Information
What happened?
(Reporting)
What is happening now?
What will happen?
(Extrapolation)
Information
Insight
How and why did it happen?
What is the next best action?
What's the best / worst that could happen?
Insight
Confidential Property of Schneider Electric | Page 45
Life Is On
Schneider Electric

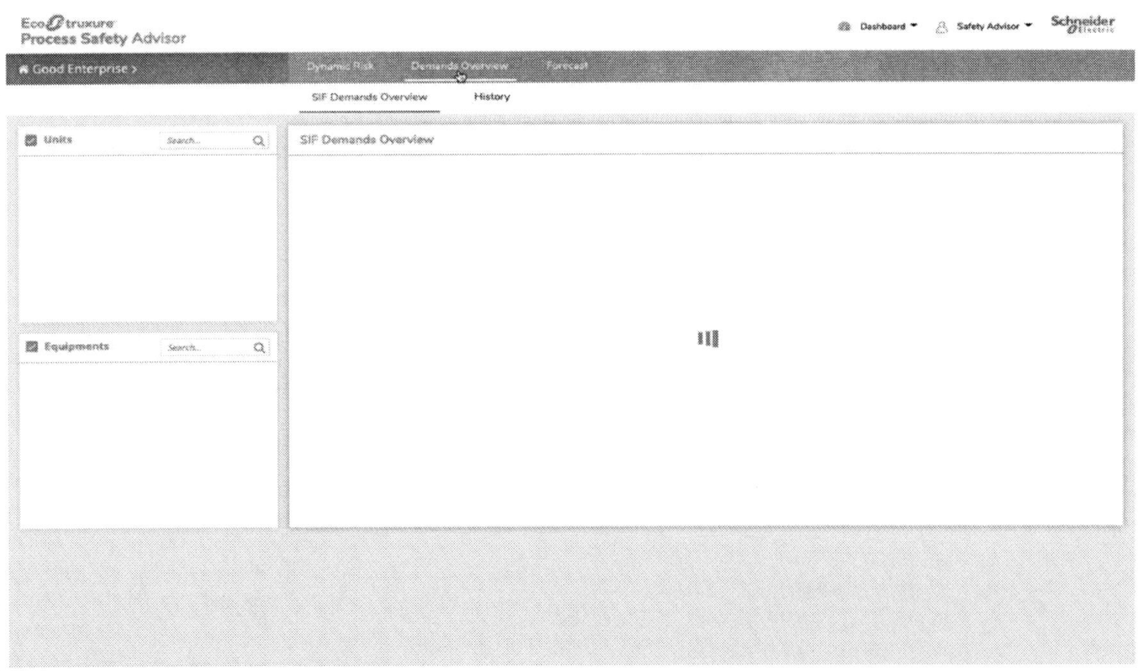
EcoStruxure
Process Safety Advisor
Dashboard
Safety Advisor
Schneider Electric
Good Enterprise >
Dynamic Risk
Demands Overview
Forecast
SIF Demands Overview
History
Units
Search...
SIF Demands Overview
Equipments
Search...

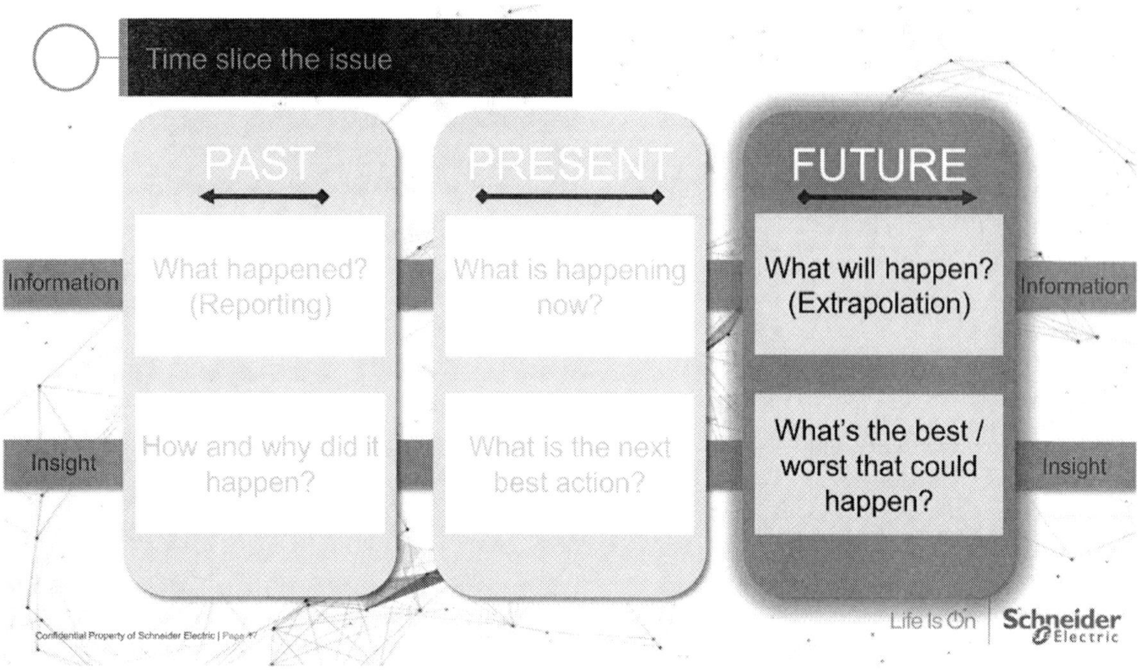
Time slice the issue
PAST
PRESENT
FUTURE
Information
What happened? (Reporting)
What is happening now?
What will happen? (Extrapolation)
Information
Insight
How and why did it happen?
What is the next best action?
What's the best / worst that could happen?
Insight
Confidential Property of Schneider Electric | Page 47
Life Is On
Schneider Electric

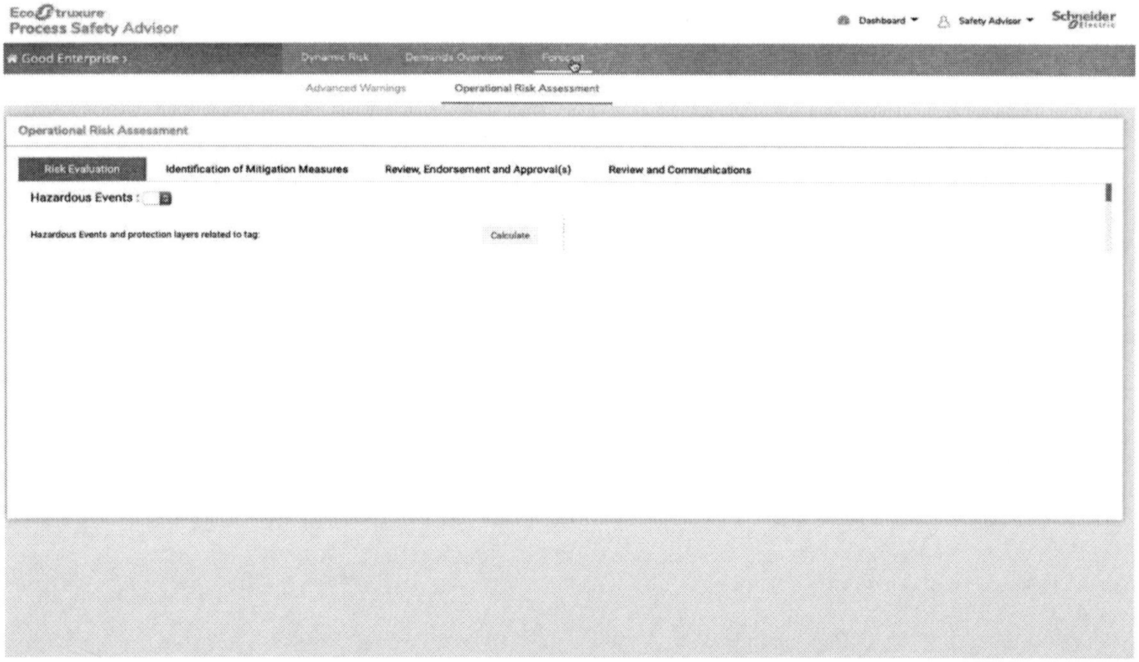
EcoStruxure Process Safety Advisor
Dashboard
Safety Advisor
Schneider Electric
Good Enterprise
Dynamic Risk
Demands Overview
Forecast
Advanced Warnings
Operational Risk Assessment
Operational Risk Assessment
Risk Evaluation
Identification of Mitigation Measures
Review, Endorsement and Approval(s)
Review and Communications
Hazardous Events :
Hazardous Events and protection layers related to tag:
Calculate

Gerd Leonhard
[Futurist]

"Science fiction is becoming Science *FACT*"
"One of the most *TRANSFORMATIONAL* times in our history"
"The change is now *EXPONENTIAL*"
"Engage with *WHAT MIGHT BE*, not just with what is"
"The new way to work is to embrace *TECHNOLOGY*"
"The future is about *HOLISTIC BUSINESS MODELS*"
"Not single improvements, but complete *TRANSFORMATIONS*"

EMBRACE TECHNOLOGY

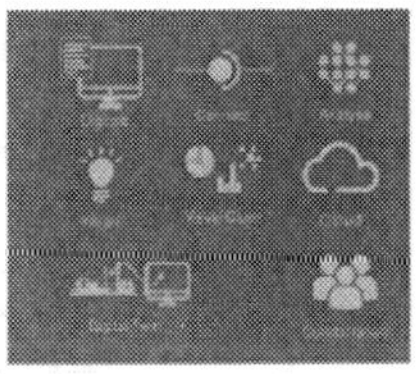

IIoT

NEW BUSINESS MODEL

DIGITAL TRANSFORMATION

TIME IS THE ENEMY

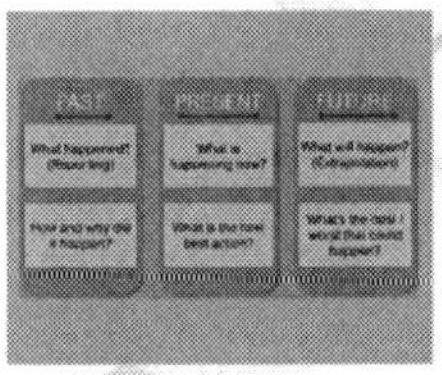

TIME SLICE THE ISSUES

FOCUS ON PEOPLE

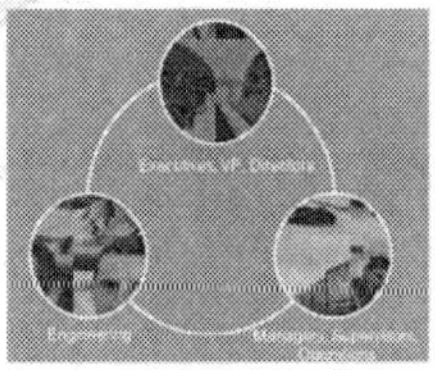

BETTER COLLABORATION

PROFIT

VALUE

Confidential Property of Schneider Electric | Page

Life Is On | Schneider Electric

Life Is On | Schneider Electric

INSIGHT IN THE USE OF DRONES AT TOTAL

2019 European Conference on Process Safety and Big Data

02/10/2019

Julien DA VELA

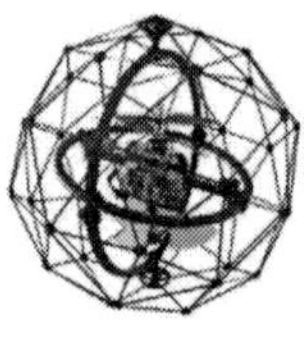

Confined spaces inspections

At height inspection

Need to be planned / not responsive
Need pilot certification
Different pilot each time

Information losses
No easy sharing of data
No easy repeatability

How do I **digitize**
gathering processes of data?

How do I **process**
data?

How do I **address**
all the other use cases?

Control & command
Data gathering

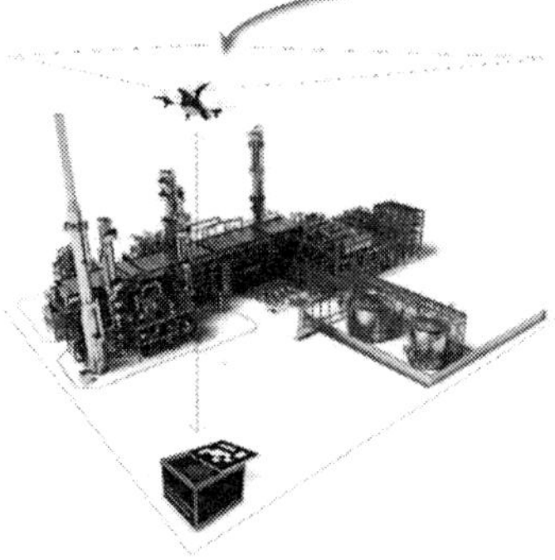

Automatic takeoff and landing
Automatic battery charge

Navigation / auto / click&go
Data in realtime and publication

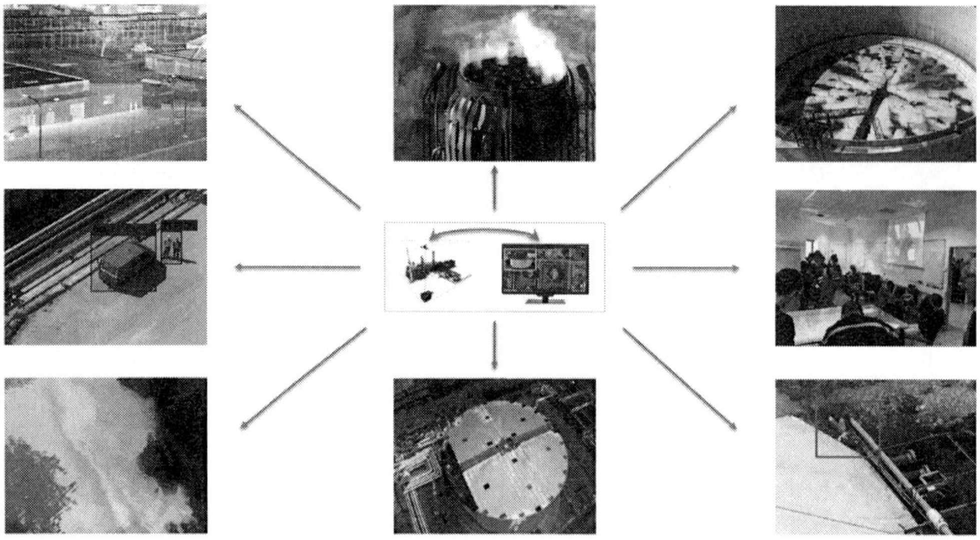

Fire exercise assisted by drone

- Accuracy of the emulsifier application
- % fire extinguishment follow up
- Ability to check if fire is well extinguished

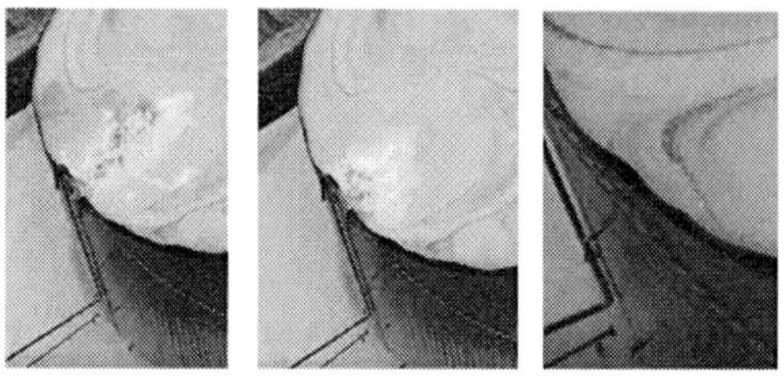

DEVELOPMENTS

- AI and Predictive models
  - People detection
  - Corrosion detection
  - Heat insulation damage detection
  - Etc.

- Sensors and gas dispersion models
  - AUSEA – GG emission
  - TADI - Multispectral camera for CH_4 detection and quantification
  - Etc.

From Risk to Resilience – avoiding failures before incidents occur

SIEMENS
Ingenuity for life

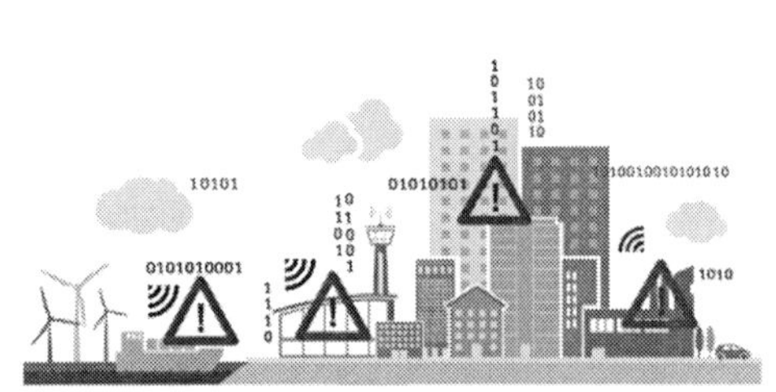

Unprotected business

- People and assets exposed to risk
- Business vulnerable to disruptions, sabotage and theft
- Costs and liability exposure
- Risk of reputational damage

Secure business

- Safer and more resilient environments and production plants
- More sustainable business, resume operations faster
- Increased operational efficiency
- Trust with customers and shareholders

Digitalization – potential to drive additional customer value

SIEMENS
Ingenuity for life

Existing products & services get digitally enhanced…

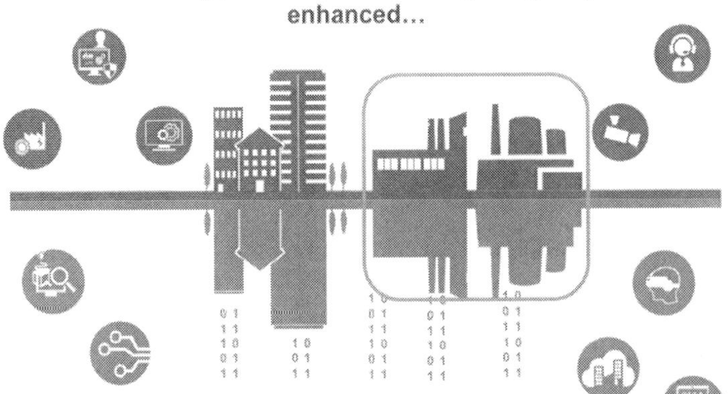

… and new data driven services are created by using data intelligently.

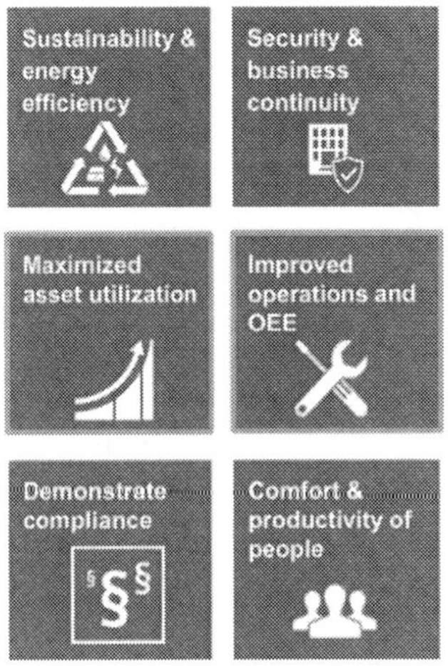

OEE = Overall Equipment Effectiveness

Value of more production vs. value of maintenance saving

SIEMENS
Ingenuity for life

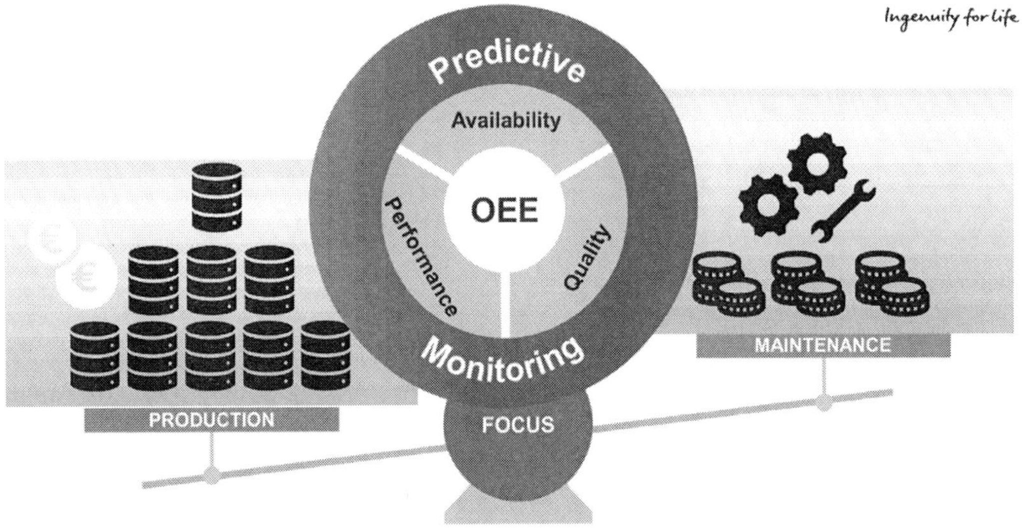

OEE = Overall Equipment Effectiveness

Equipment Predictive Analytics - Goals

SIEMENS
Ingenuity for life

Configuration and Training › **Predictive** › **Monitoring**

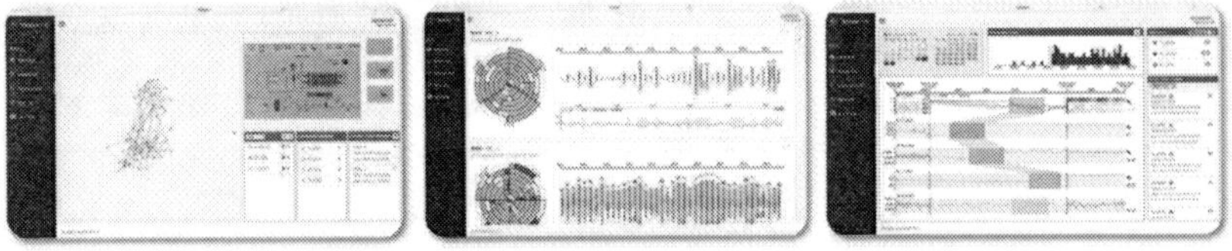

Ensure Overall Equipment Effectiveness

↑ Plant uptime	Increased plant uptime through avoidance of shut-downs by pre-alerting on failures
↑ Operation efficiency	Higher operation efficiency through predictive monitoring
↑ Decision accuracy	Better decision accuracy through identification of correlations hidden in data

- Data analysis
- Artificial intelligence
- Correlation
- Cloud-ready
- Reporting
- IT Security

Page 5 October 2019 Chemicals

SIEMENS
Ingenuity for life

What is Artificial Intelligence?

1. Historical behavior ➡ Understand the "pattern"
2. Current behavior ➡ Apply the "pattern" and raise conclusions ⬇ Create new "pattern" based on current behavior

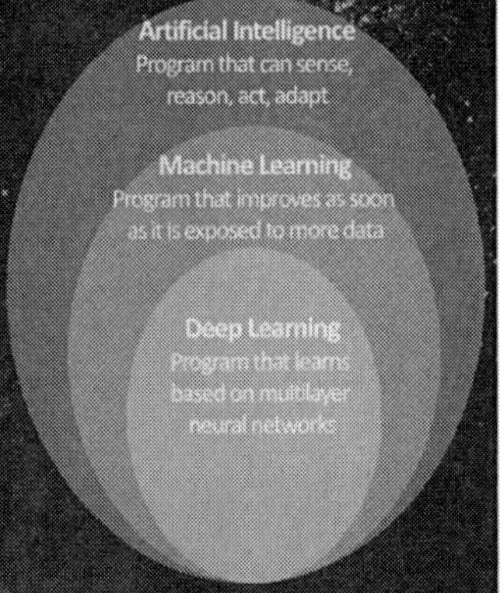

Data Availability and Analysis
On-premise or Cloud

SIEMENS
Ingenuity for life

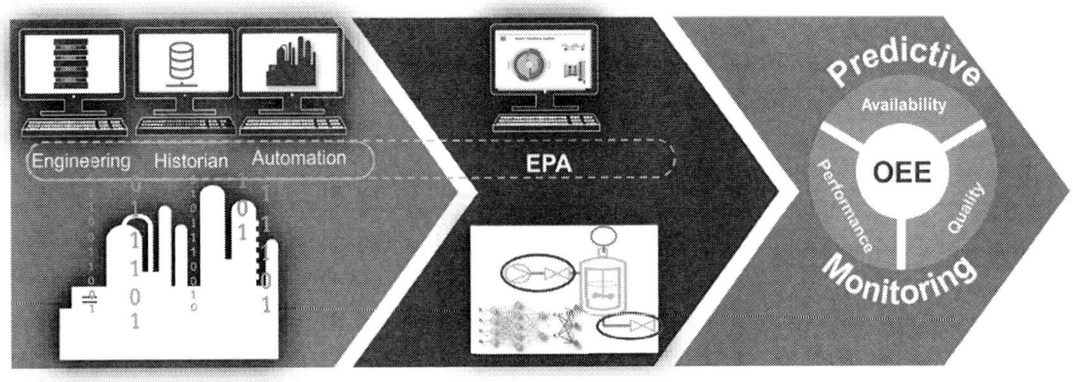

Advanced compressor monitoring – Example 1

SIEMENS
Ingenuity for life

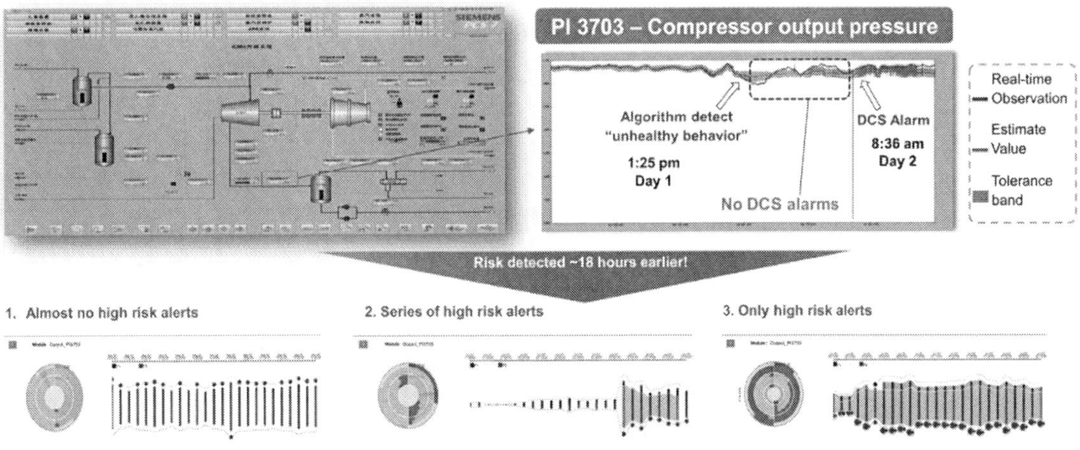

Advanced compressor monitoring – Example 2

SIEMENS
Ingenuity for life

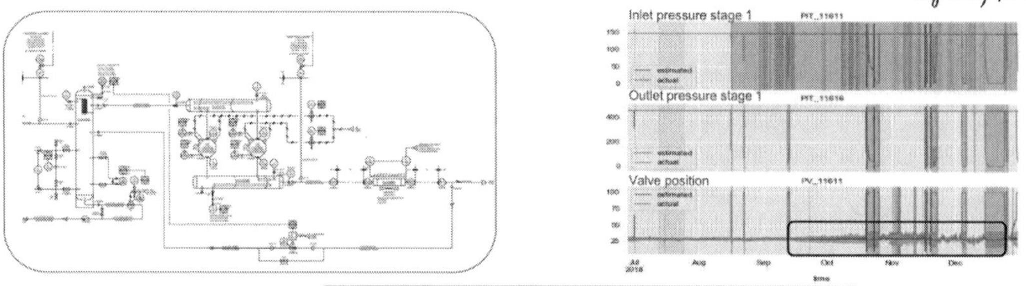

Risk detected weeks before!

1. No high risk alerts on both modules
2. High risk alerts on first module
3. High risk alerts on both modules

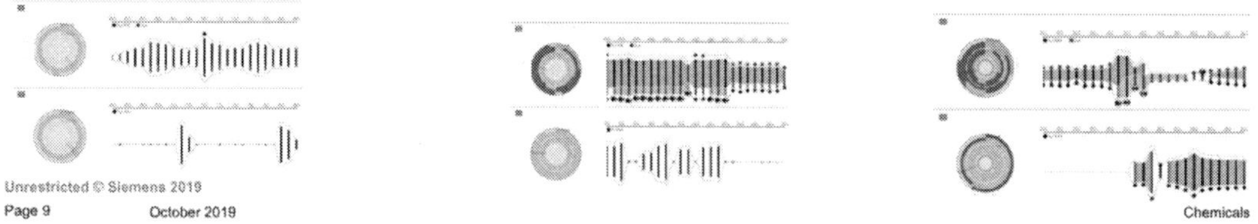

Equipment Predictive Analytics for supporting Safe Operation

SIEMENS
Ingenuity for life

Operating a process plant in a safe and reliable state is essential for high quality production and high throughput.

Critical situation may occur through faulty assets like
- Turbines
- Compressors
- Valves
- Heat exchangers

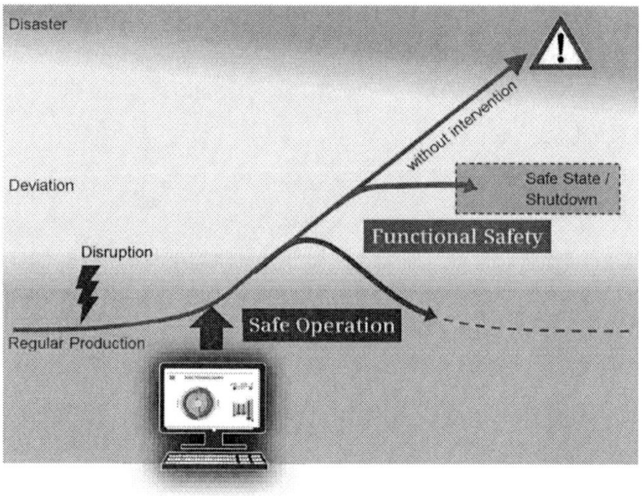

Equipment Predictive Analytics
Key Takeaways

SIEMENS
Ingenuity for life

- Run longer in safe operation.
- Rethink the process – optimize your plant
- Maximize your assets utilization
- Improve operation efficiency and OEE
- Create smart models with your data and domain know-how

Disclaimer

SIEMENS
Ingenuity for life

Subject to changes and errors. The information given in this document only contains general descriptions and/or performance features which may not always specifically reflect those described, or which may undergo modification in the course of further development of the products. The requested performance features are binding only when they are expressly agreed upon in the concluded contract.

All product designations, product names, etc. may contain trademarks or other rights of Siemens AG, its affiliated companies or third parties. Their unauthorized use may infringe the rights of the respective owner.

Thank you!

SIEMENS
Ingenuity for life

Monica Hildinger
Technology and Concepts CHEM / Glass
E-mail: [illegible]

Volker Hirsch
Technology and Concepts CHEM / Glass
E-mail: [illegible]

Siemens AG
Siemensallee 84
76187 Karlsruhe, Germany

Siemens.com/chemicals

Process Safety and Big Data 2019

Applying Deep Learning and Sequence Mining in Alarm Management

Gyula Dörgő, János Abonyi

MTA-PE Lendület Complex Systems Monitoring Research Group
Department of Process Engineering, University of Pannonia

The alarm management lifecycle

Which alarms are necessary?

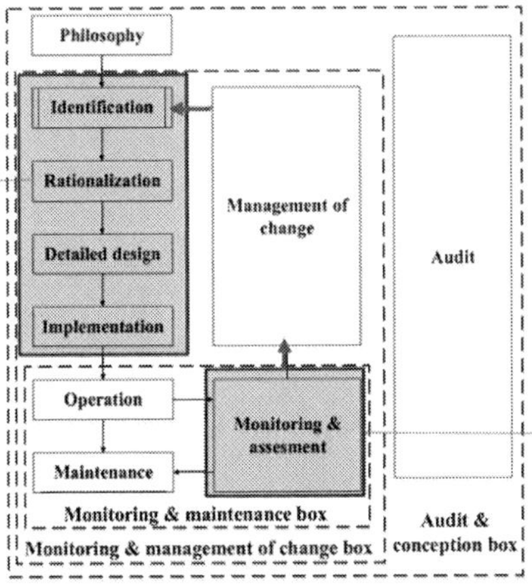

Measuring performance, KPIs

ANSI/ISA – 18.2
Management of Alarm Systems for the Process Industries

1/18

Big data and ML can support the alarm management?

From	To	Production Unit	Unit	Tag	Event Type	Description
10/24/2018 16:02	10/24/2018 16:04	Distillation	Main column	Head temp.	Alarm	High alarm
10/24/2018 16:02	10/24/2018 16:04	Distillation	Main column	Cond. cooling	Operator action	Open

Millions of data points

Data (process) mining, machine (deep) learning...

Something useful ?

Delayed coker plant at MOL:

Number of tags: ~8000
Number of units: 19
Number of events:
~2Milion / falf year

2/18

The concept of data-driven alarm management

PAST

FUTURE

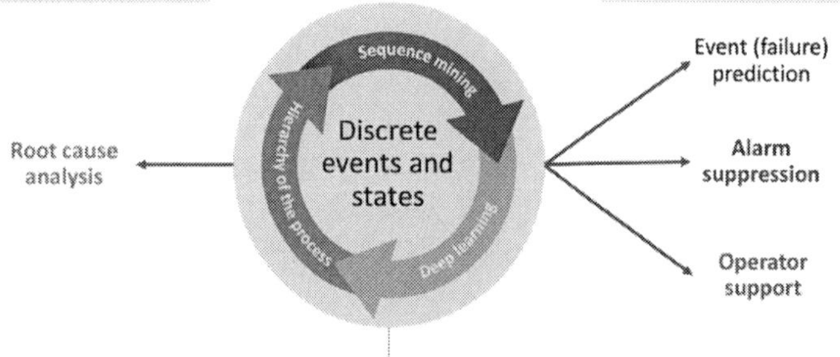

Discrete events
- State of the process
- Frequent sequence mining for prediction
- Alarm – operator interactions

Deep learning
- Root cause analysis
- Event prediction
- Visualisation for interpretability

Hierarchy of the process
- Structural process models
- Helps the freq. seq. mining
- Monitoring the spillover of malfunctions

3/18

Illustration of the frequent pattern-based concept

ALARMS

Root cause analysis: 60 % confidence

Grouping: Parent-children

Prediction: 80 % of the cases

$$Fouling \Rightarrow F^{L}_{reflux} \Rightarrow p^{H}_{condenser} \Rightarrow T^{H}_{column\ top} \Rightarrow T^{H}_{condense} \Rightarrow A^{L}_{product}$$

OPERATOR ACTIONS

Automation

Recommendation

$$Flow^{H}_{pump} \Rightarrow Q^{H}_{cooling} \Rightarrow Flow^{L}_{inflow} \Rightarrow Bypass$$

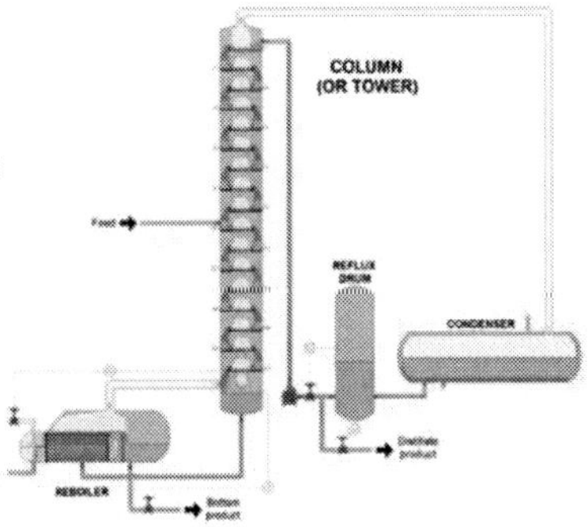

4/18

The method and the framework

Open source machine learing tools are applicable in the process industry

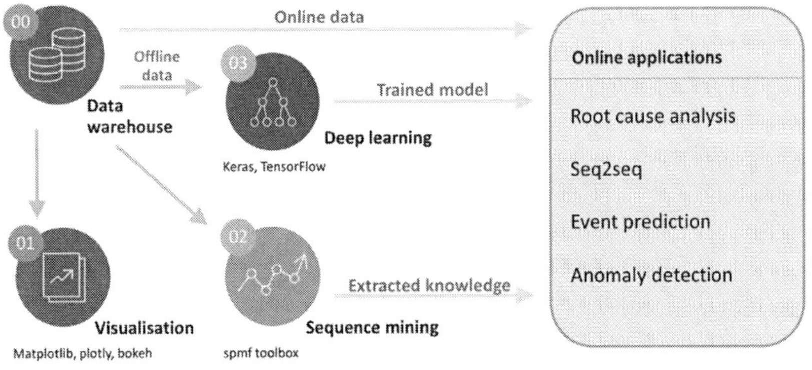

python

Keras

ProM
process mining workbench

5/18

Visualisation of the operator workload

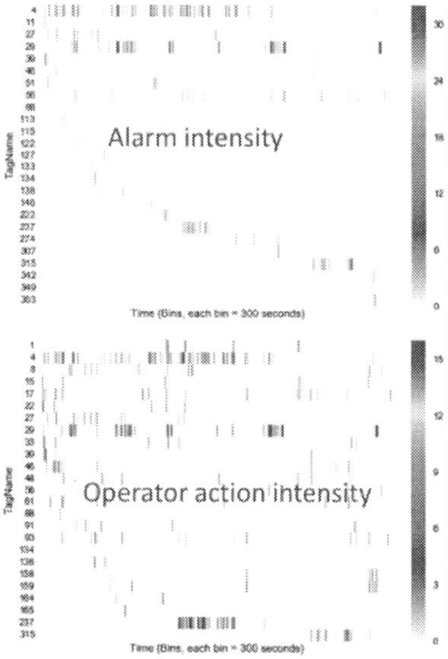

Alarm performance metrics based on 30< days of data		
Metric	Target Value	
	Acceptable	Max.
Annunciated alarms per hour per operator console	~6	~12
Annunciated alarms per 10 minutes per operator console	~1	~2
Percentage of 10-minute periods containing more than 10 alarms	~<1 %	
Maximum number of alarms in a 10-minute period	≤10	
Percentage of time the alarm system is in a flood condition	~<1 %	
Percentage of the top 10 most frequent alarms to the overall alarm load	~<1 % to 5 % maximum	
Quantity of chattering and fleeting alarms	Zero	
Stale (standing) alarms	Less than 5 present on any day	
Annunciated priority distribution 3 priorities: (Low-Medium-High)	80 – 15 – 5 %	

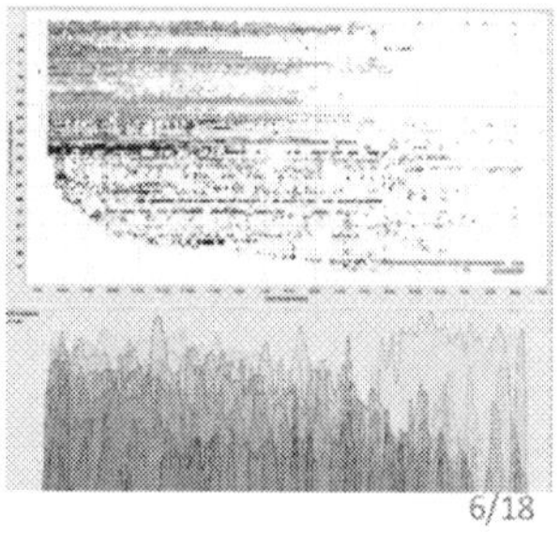

6/18

Multi temporal pattern representation

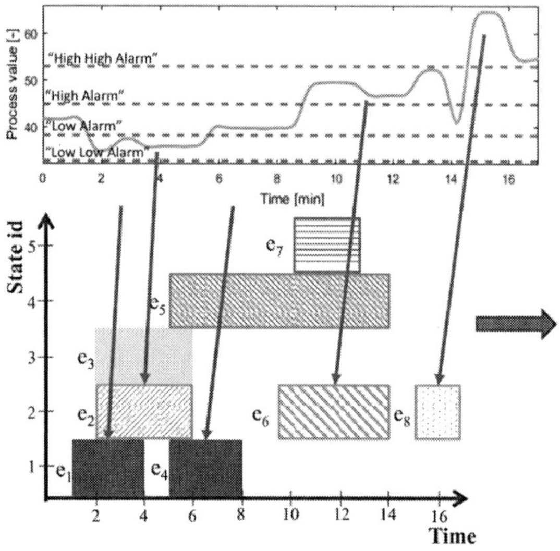

e event $- < s, st, et >$

$e_1 \overset{O}{\Rightarrow} e_2,\ e_2 \overset{E}{\Rightarrow} e_3,\ e_6 \overset{D}{\Rightarrow} e_7$

Event id	State id	Start Time	End Time
e_1	s_1	1	4
e_2	s_2	2	6
e_3	s_3	2	6
e_4	s_1	5	8
e_5	s_4	5	14
e_6	s_2	10	14
e_7	s_5	11	13
e_8	s_2	16	17

7/18

The probability and the prediction power of a sequence

$$\phi_k := s_0 \overset{R_1}{\Rightarrow} s_1 \overset{R_2}{\Rightarrow} s_2 \overset{R_3}{\Rightarrow} \dots \overset{R_k}{\Rightarrow} s_k$$

The probability of a sequence is calculated based on the number/frequency of occurence (called: *support*)

$$P(s_i) \sim supp(s_i)$$

The probability of a transition (called: confidence)

$$P(s_k|\phi_{k-1}) = \frac{P(\phi_k)}{P(\phi_{k-1})} = \frac{supp(\phi_k)}{supp(\phi_{k-1})}$$

SID	Sequence
1	$a_1 \Rightarrow a_2 \Rightarrow a_4 \Rightarrow o_1 \Rightarrow o_2$
2	$a_1 \Rightarrow a_2 \Rightarrow o_1$
3	$a_2 \Rightarrow a_3 \Rightarrow o_3$
4	$a_5 \Rightarrow o_1$
5	$a_1 \Rightarrow a_2 \Rightarrow a_4$
6	$a_1 \Rightarrow a_2 \Rightarrow a_4 \Rightarrow o_1$

FSID	Sequence	Support	Confidence
1	a_1	2/3	1
2	a_2	5/6	1
3	a_4	1/2	1
4	o_1	2/3	1
5	$a_1 \Rightarrow a_2$	2/3	1
6	$a_1 \Rightarrow a_2 \Rightarrow o_1$	1/2	3/4
7	$a_1 \Rightarrow a_2 \Rightarrow a_4$	1/2	3/4

8/18

Frequent sequence mining of alarm–operator action series

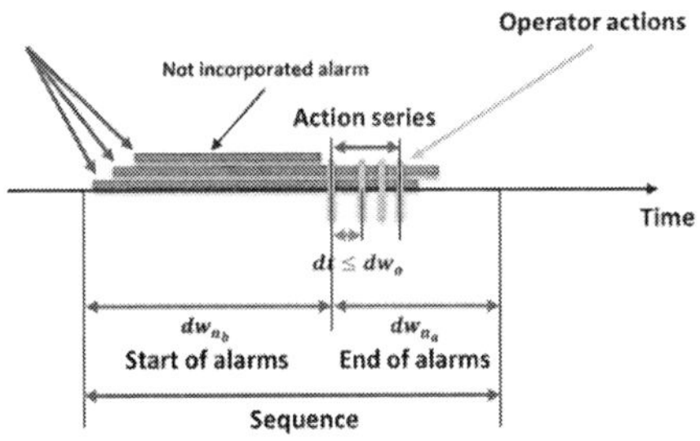

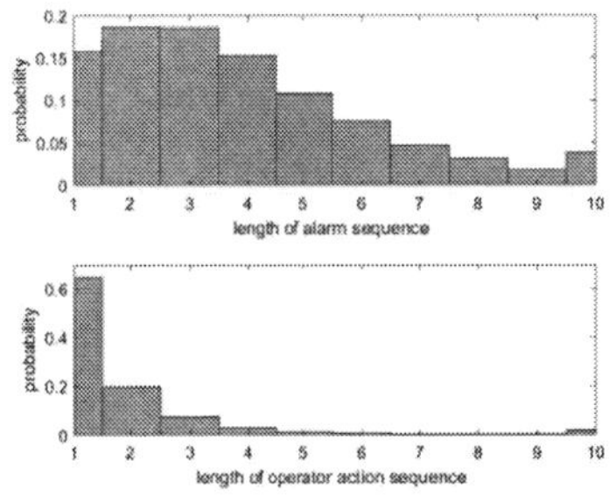

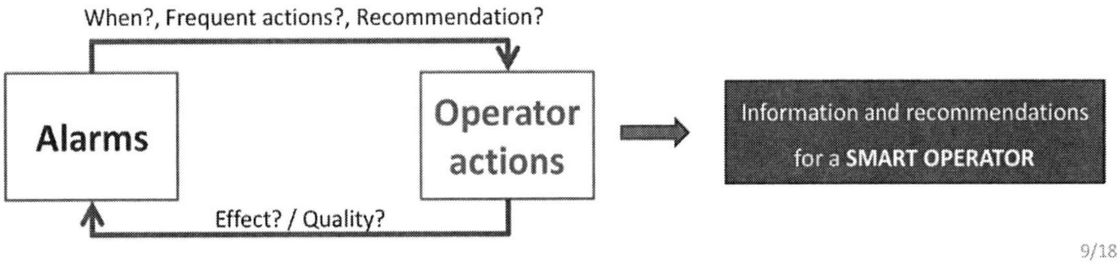

9/18

Sequences at the hierarchical levels

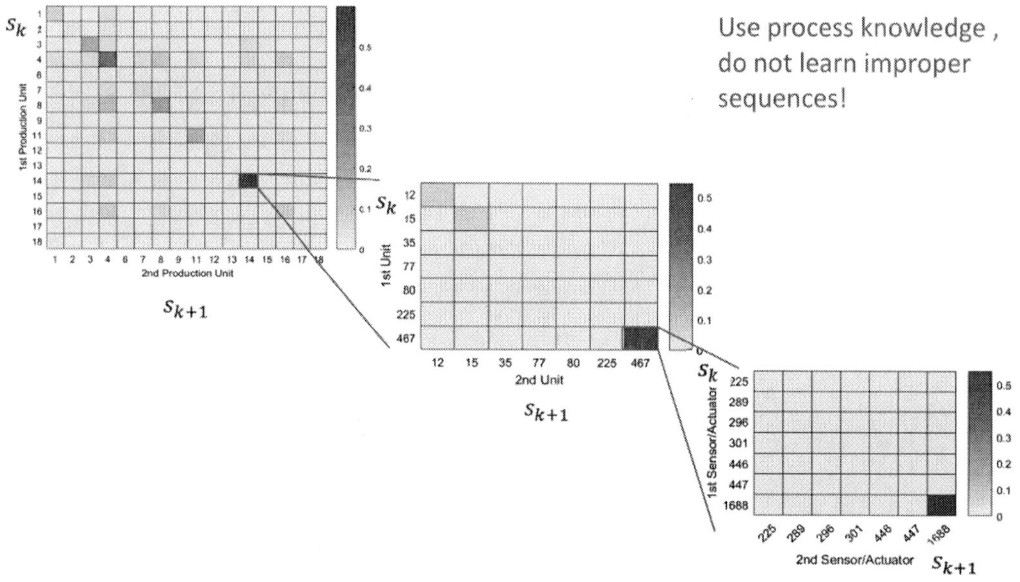

Use process knowledge , do not learn improper sequences!

10/18

Spillover effect of malfunctions

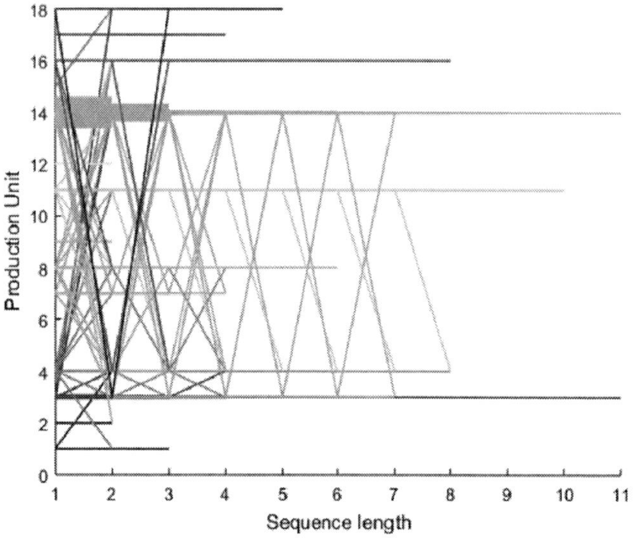

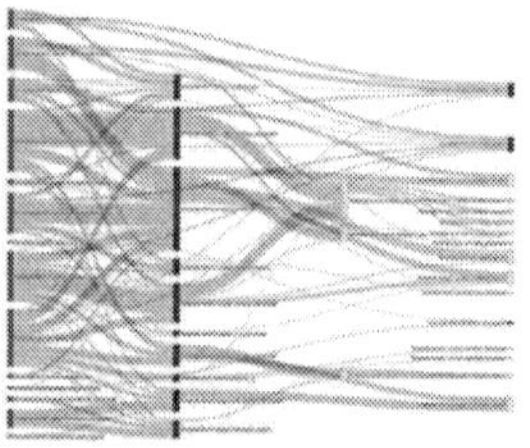

ID	unit	unitname
1	AS	DCU AMINE REGENERALO
2	D1	DCU RAW MATERIAL SYSTEM
3	D2	DCU H-101 FURNACE
4	D3	DCU COKE TANKS
5	D4	DCU MAIN FRACTIONATOR
6	D5	DCU LCGO HCGO PRODUCT SYST.
7	D6	DCU BLOWDOWN SYST.
8	FU	FLARE SYSTEM
9	G1	DCU MOIST. GAS SYSTEM
10	G2	DCU HEATING GAS LINE
11	G3	DCU C3/C4 SEPARATION
12	G4	DCU C3/C3= SEPARATION
13	G5	DCU BENZIN SEPARATION
14	S1	DCU UTILITY SYSTEMS 1
15	S2	DCU UTILITY SYSTEMS 2
16	S3	DCU UTILITY SYSTEMS 3
17	S4	DCU UTILITY SYSTEMS 4
18	AC	ADVANCED CONTROL
19	C3	CLAUS UNIT

11/18

The tools of network science are applicable for the evaluation of risks, support the operators, ...

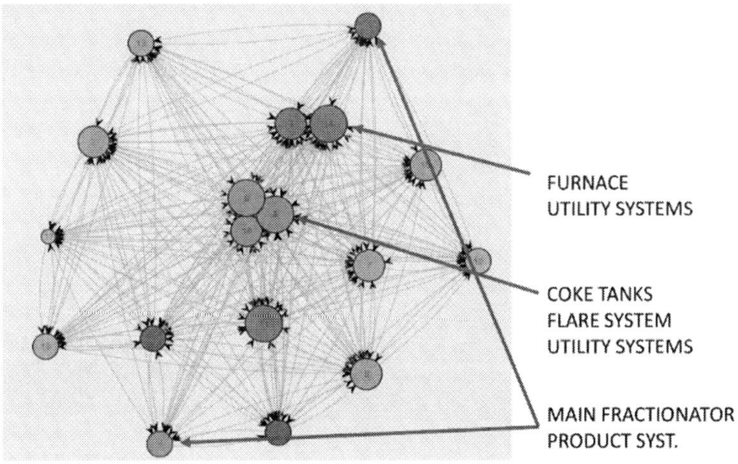

Circles represent units
Closer – more connected
Bigger– bigger importance

ID	unit	unitname
1	AS	DCU AMINE REGENERALO
2	D1	DCU RAW MATERIAL SYSTEM
3	D2	DCU H-101 FURNACE
4	D3	DCU COKE TANKS
5	D4	DCU MAIN FRACTIONATOR
6	D5	DCU LCGO HCGO PRODUCT SYST.
7	D6	DCU BLOWDOWN SYST.
8	FU	FLARE SYSTEM
9	G1	DCU MOIST. GAS SYSTEM
10	G2	DCU HEATING GAS LINE
11	G3	DCU C3/C4 SEPARATION
12	G4	DCU C3/C3= SEPARATION
13	G5	DCU BENZIN SEPARATION
14	S1	DCU UTILITY SYSTEMS 1
15	S2	DCU UTILITY SYSTEMS 2
16	S3	DCU UTILITY SYSTEMS 3
17	S4	DCU UTILITY SYSTEMS 4
18	AC	ADVANCED CONTROL
19	C3	CLAUS UNIT

12/18

Deep learning (LSTM models) are applicable for root cause analysis

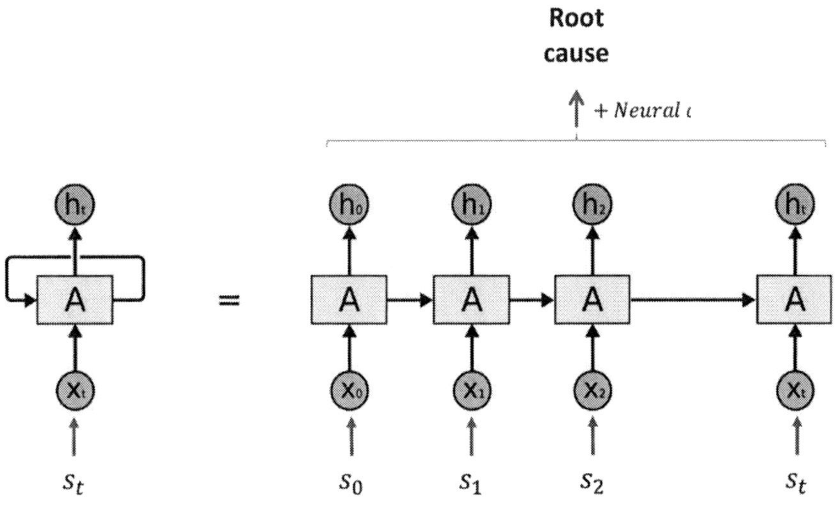

13/18

The PCA of the embedding layer is applicable for alarm grouping

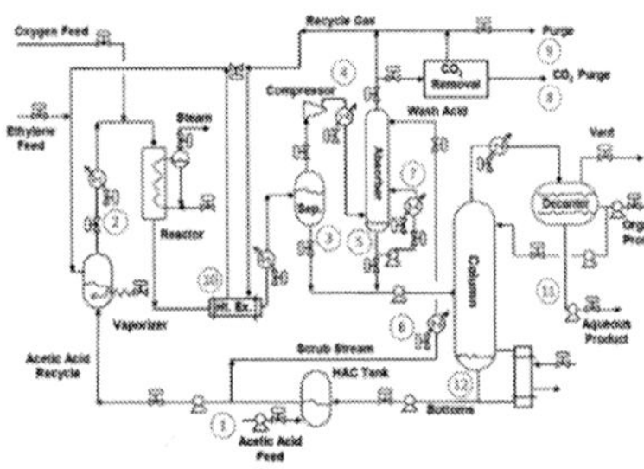

Event	Name of event	Faults
145	Circulation Stream Temperature	4, 6
245	Decanter Aqueous Level	11
244	Decanter Aqueous Level	1, 2, 3, 4, 5, 6, 7, 8,9, 10, 11
85	Separator Level	3, 9
195	FEHE Hot Exit Temperature	9
64	Heater Exit Temperature	2, 9
194	FEHE Hot Exit Temperature	4, 6, 7, 9
71	Reactor Exit Temperature	1, 2, 3, 4, 5, 6, 7, 8, 9, 10, 11
61	Heater Exit Temperature	2
161	Scrub Stream Temperature	1, 7
165	Scrub Stream Temperature	1, 3, 7
235	Decanter Organic Level	10
31	HAc Tank Level	1
114	Compressor Exit Temperature	2, 4, 9
115	Compressor Exit Temperature	4
125	Absorber Level	1, 2, 3, 4, 5, 6, 7, 9, 11

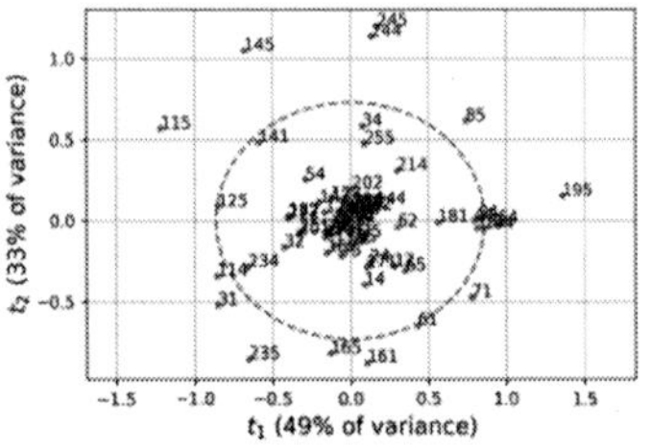

14/18

Deep learning is applicable for sequence prediction

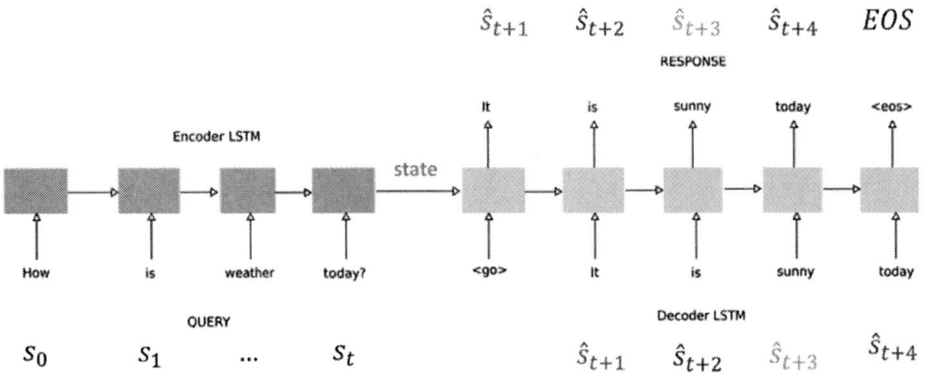

15/18

Sequence prediction is applicable for alarm suppression

Seq2seq DL- VAc

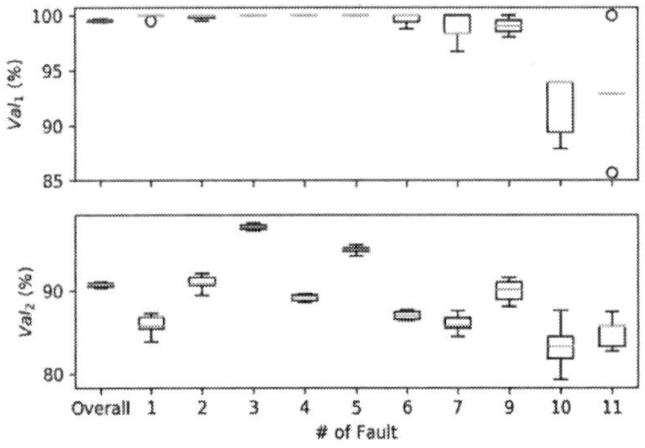

Freq seq- VAc

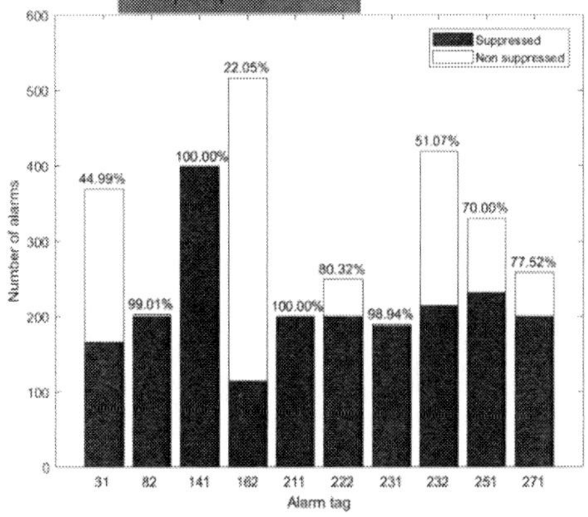

$$\phi_k = 82 \overset{B}{\Rightarrow} 191 \overset{O}{\Rightarrow} 41 \overset{E}{\Rightarrow} \ldots \overset{B}{\Rightarrow} 31$$

16/18

Knowledge Discovery can support alarm management

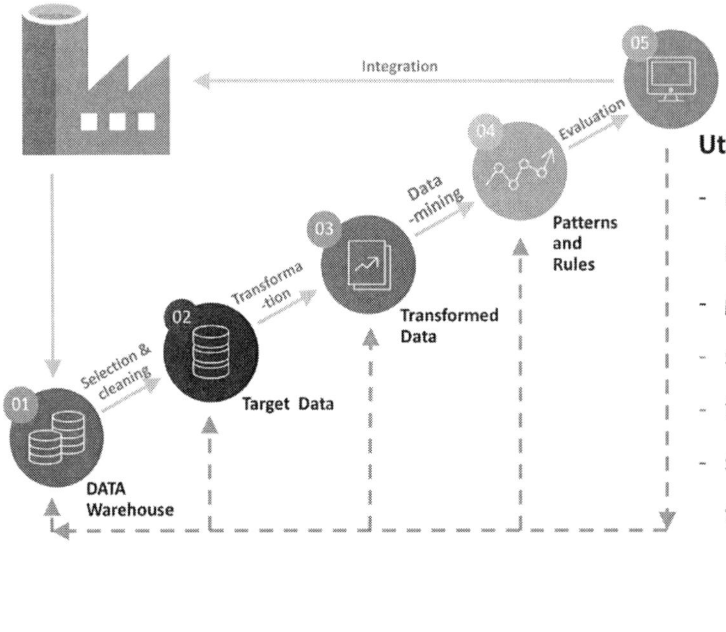

Utilisation of the knowledge

- monitoring of alarm rationalisation projects
- grouping alarms
- suggest alarm supression rules
- fault detection and diagnosis
- support and monitor of the operators

17/18

Conclusions

The related ML task

- Sequence mining
- Event prediction
- Root cause analysis

Problem

Realization

- Visualisation
- Advanced techniques
- **Incorporation of process-relevant knowledge**
- Open source

Solution

Future

- KPIs of operation actions
- Rare events?
- Recommendation system

Improvement

The python script-based solutions are well integrable

www.abonyilab.com

18/18

References

[1] Gyula Dörgő, János Abonyi: Sequence Mining based Alarm Suppression, *IEEE Access*, Vol. 6(1) pp. 15365-15379 (2018), DOI: 10.1109/ACCESS.2018.2797247

[2] Gyula Dörgő, Kristóf Varga, Máté Haragovics, Tibor Szabó, János Abonyi: Towards Operator 4.0, increasing production efficiency and reducing operator workload by process mining of alarm data, *Chemical Engineering Transactions*, Vol. 70 pp. 829–834 (2018), DOI: 10.3303/CET1870139

[3] Gyula Dörgő, János Abonyi: Learning and predicting operation strategies by sequence mining and deep learning, submitted to *Computers & Chemical Engineering*, Under revision

[4] Gyula Dörgő, Péter Pigler, János Abonyi: Understanding the importance of process alarms based on the analysis of deep recurrent neural networks trained for fault isolation, *Journal of Chemometrics*, Vol. 32(4) pp. e3006 (2018), DOI: 10.1002/cem.3006

[5] Gyula Dörgő, Péter Pigler, Máté Haragovics, János Abonyi: Learning operation strategies from alarm management systems by temporal pattern mining and deep learning, Proceedings of the 28th European Symposium on Computer Aided Process Engineering, (ESCAPE), in *Computer Aided Chemical Engineering* book series, vol. 43, pp. 1003-1008, 2018.

[6] Gyula Dörgő, Kristóf Varga, János Abonyi: Hierarchical frequent sequence mining algorithm for the analysis of alarm cascades in chemical processes, *IEEE Access*, Vol. 6 pp. 50197-50216 (2018), DOI: 10.1109/ACCESS.2018.2868415

19/18

2019 EUROPEAN CONFERENCE
ON PROCESS SAFETY & BIG DATA
CCPS EPSC
PRESENTS THE
WORLDWIDE PREMIERE OF
THE STARCHECK TEAM
FACES THE CHALLENGE
TO SAVE PLANET AkzBASTATDS
FROM GRAVE IMMINENT DANGER
WITHOUT DYNAMIC
CONNECTIONS
YOUR EXISTING
HSE COMPLIANCE
TOOLS
ARE FUTILE
STARCHECK PART II
DYNAMIC CHECKLISTS
VERSATEC ENERGY B.V. & RADIAL SOFTWARE GROUP B.V. PICTURES PRESENT AN IT'S NOT ROCKET SCIENCE PRODUCTION
STARCHECK PART II: DYNAMIC CHECKLISTS STARRING MARK VAN 'T ZET/RADIAL SG AS CAPTAIN MARK
JOEP COENEN/VERSATEC AS COMMANDER JOEP KLAAS-OTTO YKEMA/VERSATEC AS Mr. B.O. TIE
film editor PETER BRINK/RADIAL SG casting by WWW.VERSATEC.NL / WWW.RADIALSG.COM
SPECTRAL RECORDING
DOLBY STEREO SR
IN SELECTED THEATRES

StarCheck Part II: Dynamic Checklists

IMDB Rating: 9.6/10

Plot Summary

Join the StarCheck crew as they face their biggest challenge yet: saving the planet Akzbastatds from grave imminent danger.

Having cleared the space debris at EPSC2018, Captain Mark and Commander Joep steer their HSE Enterprise spaceship towards an uncharted piece of the galaxy in their mission to explore unknown hazardous worlds.

Upon arrival at the industrial planet Akzbastatds, they soon find out that the HSE Management System is missing crucial features, reminding them of the process industry back on Earth as it was 50 years ago. They explain their concept of a Dynamic Checklist to the operations manager of Akzbastatds, mr. B.O. Tie, and display the ways in which the current tooling is failing on the ship's viewport.

Soon, the crew of HSE Enterprise end up in an unexpected emergency situation... Can they save the day using their extensive industry experience and advanced tooling?

Find out at www.versatec.nl / www.viewport.ai

Value of BIG data for instrumentation in brownfield plants for process safety and reliability

Value proposition

- What do we need to get more data?
  - We have smart instruments, but we do not use them in this way.
- Reliability:
  - How many unplanned events in our plants
- Safety:
  - Do not test so often;
  - exchange equipment based on manufacturers manuals

10 10 10 BY2020 Technical Expertise & Support Leverage Globally Act Regionally Execute Locally *Faster and Smarter*

What equipment do we find in our plants?

„simple" equipment (no health monitoring possible)	„intelligent" equipment (with health monitoring possibility)
pumps	transmitters (pressure; temperature; flow, ...)
compressors	control valves
manual valves	
on/off valves automated	
vessels / tanks	
piping	
......	

External measurement systems have to be installed to get diagnostic data.

We have to install systems to get the data out of the equipment.

What data are we talking about?

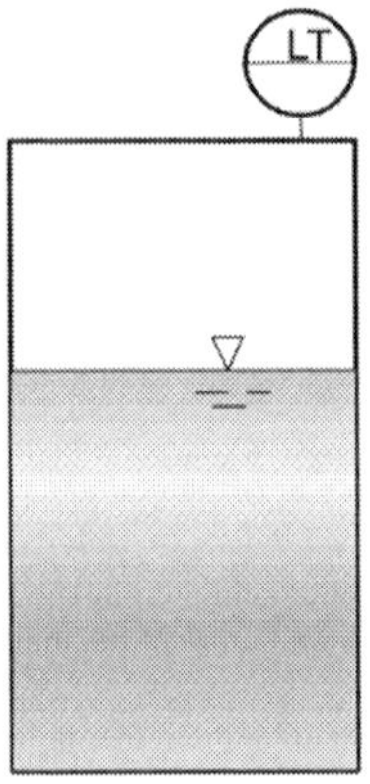

	Measurement	Use of the signal	Signal
primary function	level measurement	level control for operations	4-20 mA
secondary function	realizes foam	process upset (parameter)	HART (FDT/DTM DDL)
tertiary information	information of the internal condition	early warning of upset	HART (FDT/DTM DDL)
static Information	set of parameters	simplified exchange strategy	HART (FDT/DTM DDL)
static Information	information about the device specification	simplified maintenance strategy	HART (FDT/DTM DDL)

Where is the right place to get the HART information?

at the instrument (wireless HART)	in the junction box (wireless/wired HART)	in the cross board room (wireless/wired Hart)	in the control system (maintenance work station)

Picture taken from: http://www.smar.com/en/product/tt400w-wirelesshart-temperature-transmitter

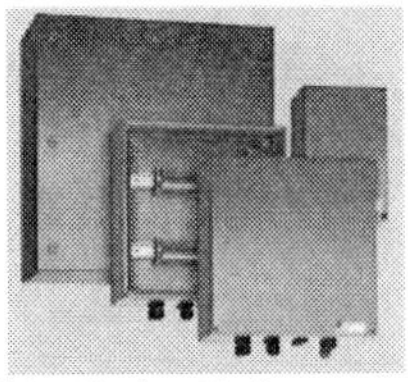

Picture taken from: https://r-stahl.com/en/global/products/junction-and-terminal-boxes/terminal-boxes-industrial-version/terminal-boxes-series-g150-39699/259735-39713/

Picture taken from: https://www.wago.com/global/installation-terminal-blocks-and-connectors/topjob-s-fi-main-distribution-board-set/p/821-122

OS | MS | ES

DCS

ES_ Engineering Station

OS_ Operator Station

MS_ Maintenance Station

DCS_ Distributed control system

PIC — JB-xxxxx — Terminal-xxxxx

10 10 10 BY2020 — Technical Expertise & Support — Leverage Globally — Act Regionally — Execute Locally — *Faster and Smarter*

DOW

Reliability

First approach: product loss

- What is the percentage of instrument that causes a serious plant trip per year? (False Trip Frequency)

< 0,1% | </= 1 % | > 1 %

Second approach: test effort

What is the percentage of instruments that are part of a regular test program?

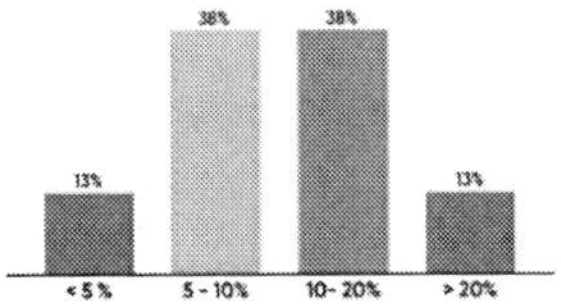

What is the percentage of the services that are legally required

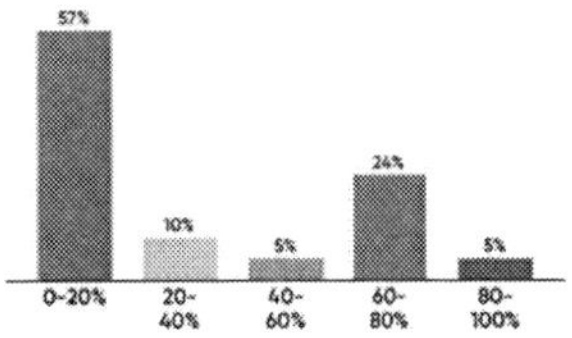

Can we justify putting money and work force into a program?

10 10 10 BY2020 — Technical Expertise & Support — Leverage Globally — Act Regionally — Execute Locally — *Faster and Smarter*

DOW

Reliability

First approach: product loss

- How many instrument related problems have caused a serious plant trip?
  - 1 out of 55.000 instruments per year << 0,1%

Second approach: test effort

- What is the percentage of instruments that are part of a regular test program?
  - 15 %
- How many of these tests are legally required?
  - 90 %
- → Do we do this for 1,5% of the instruments where we have reliability reasons?

(By the way, these are to 95% analytical devices pH, conductivity, O2....)

Can we justify putting money and work force into a program?

10 10 10 BY2020 Technical Expertise & Support Leverage Globally Act Regionally Execute Locally *Faster and Smarter*

Safety (functional safety)

$$PFD_{avg\,1oo1} = \frac{1}{2} \cdot \boxed{\lambda_{DU}} \cdot DC_{test} \cdot T_{test} + \frac{1}{2} \cdot \boxed{\lambda_{DU}} \cdot (1 - DC_{test}) \cdot \boxed{T_{eol}}$$

What is the typical test interval we should use for our safety application?

With today's manufacturer data we can calculate test intervals of 60 months or more for a 1oo1 configuration (for SIL 1, and 3 years for SIL 2).

For redundancy (1oo2) more than 6 years for SIL 2 and to 6 years for SIL 3 are possible.

Do we need more?

What is the typical exchange age we should use for our equipment in safety applications?

Normally, the manufacturers can guarantee a constant failure rate of ~10 years for the PFD calculations. We often use this devices for around 20 years.

Here it makes sense to generate data.

10 10 10 BY2020 Technical Expertise & Support Leverage Globally Act Regionally Execute Locally *Faster and Smarter*

DOW CONFIDENTIAL - Do not share without permission

Safety (functional safety) Partial test

$$PFD_{avg} = \frac{1}{2} \cdot \lambda_{DU} \cdot \boxed{DC_{test1} \cdot T_{test1}} + \frac{1}{2} \cdot \lambda_{DU} \cdot (DC_{test2} - DC_{test\,1}) \cdot T_{test\,2} + \frac{1}{2} \cdot \lambda_{DU} \cdot (1 - DC_{test2}) \cdot T_{eol}$$

Where does inline (partial) test make sense?

It makes sense for flow, valves and some level devices. For flow many companies have devices that can check the electronics to a high degree. Valves in safety applications are typically on/off, so there is no positioner. Partial stroke test is a limited option for pneumatic valves.

<u>Here is a real value case I see:</u>

- **Testing of inline flow devices.**
- **Testing of level devices (without pulling them)**
- **Solenoid that records torque (pressure) reading during the last turn when the the plant goes into shut down.**
- **Extend the end of life.**

Functional Safety: to put this into perspective

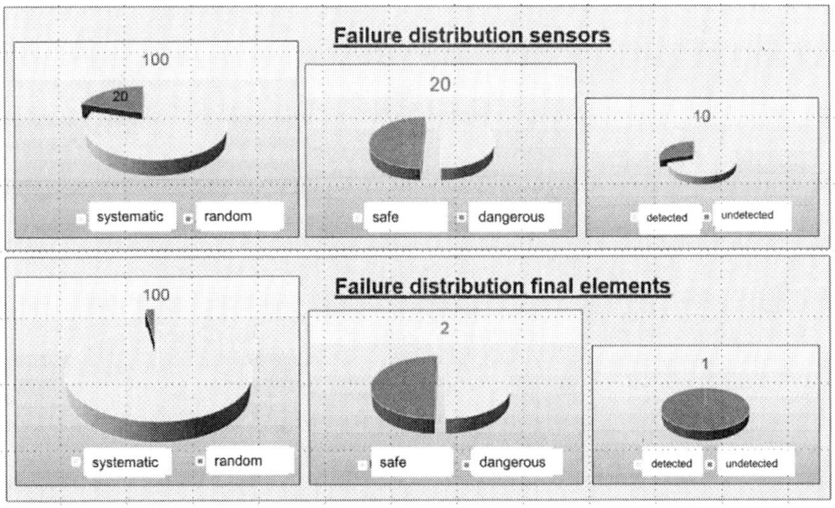

Thank you very much for your attention.

If you have further questions, please contact me:

cdemski@dow.com

Let's start the discussion!

Christian Demski
Dow Deutschland Anlagenge...
Discipline Activity Leader
Technical Expertise & Support
+49 41469 13814 Work
+49 1523 8255298 Other
CDemski@dow.com
Postfach 1120
Stade, ND 21677

10 10 10 BY2020

Technical Expertise & Support
Leverage Globally
Act Regionally
Execute Locally
Faster and Smarter

AUTHOR INDEX

AIChE
120 Wall Street, FL 23
New York, NY 10005-4020

ISBN 978-1-7138-2864-8

CURRAN ASSOCIATES INC.
proceedings
.com